low to Solve Math Word Problems on Standardized Tests

David S. Wayne, Ph.D.

McGraw-Hill

New York   Chicago   San Francisco   Lisbon   London
Madrid   Mexico City   Milan   New Delhi
San Juan   Seoul   Singapore
Sydney   Toronto

Library of Congress Cataloging-in-Publication Data

Wayne, David S.
How to solve math word problems on standardized tests / David Wayne.
p. cm.
Includes index.
ISBN 0-07-137693-3 (pbk. : acid-free paper)
1. Mathematics—Problems, exercises, etc.   2. Word problems (Mathematics)
I. Title.

QA43 .W36   2002
510′.76—dc21   2001054603

McGraw-Hill

A Division of The **McGraw·Hill** Companies

1 2 3 4 5 6 7 8 9 0   DOC/DOC   0 7 6 5 4 3 2 1

ISBN 0-07-137693-3

*The sponsoring editor for this book was Barbara Gilson, the editing supervisor was
Ruth W. Mannino, and the production supervisor was Maureen Harper. It was set in
Stone Serif by TechBooks.*

Printed and bound by R. R. Donnelley and Sons Company.

McGraw-Hill books are available at special quantity discounts to use as premiums
and sales promotions, or for use in corporate training programs. For more
information, please write to the Director of Special Sales, Professional Publishing,
McGraw-Hill, Two Penn Plaza, New York, NY 10121-2298. Or contact your
local bookstore.

This book is printed on recycled, acid-free paper containing
a minimum of 50% recycled de-inked fiber.

Contents

Preface   v

Chapter 1—Making Use of Multiple Choices
    Problems whose solutions can be found
    by working backward from the choices   1

Chapter 2—Working with Many Variables
    Problems whose solutions can be found
    by substituting numbers into the problems   21

Chapter 3—Using Ratios and Proportions Quickly
    Problems whose solutions can be found
    by applying simple ratios and proportions   47

Chapter 4—Using Lists, Patterns, and Diagrams
    Problems whose solutions can be found
    by creating lists and looking for patterns   73

Chapter 5—Problems Asking for a Comparison
    Problems from all areas in the two-column
    quantitative comparison format   104

Chapter 6—Additional Problems for Practice
    A set of miscellaneous problems
    for additional practice   118

Table of Common Measurements and Conversions   **128**

Table of Common Formulas Seen in Word Problems   **130**

Table of Fundamental Geometric and Trigonometric
   Relationships   **133**

Appendix A—A Brief Review of Fractions   **137**

Appendix B—A Brief Review of Signed Numbers   **141**

Appendix C—A Brief Review of Exponents   **145**

Appendix D—A Brief Review of Radicals   **148**

Appendix E—A Brief Review of Polynomials   **151**

Appendix F—A Brief Review of Solving Equations   **156**

Index   **167**

iv

Preface

A standardized test attempts to measure knowledge and ability by having the test taker answer questions that are representative of a common set of facts and understandings. These tests are scored in a way that ranks the test taker against the entire population. Standardized tests are used as instruments to measure student performance, as entrance exams for schools, and as qualifying exams for professional certification. Examples include statewide public school assessments, diploma equivalency exams such as the GED, college entrance exams such as the PSAT, SAT, and ACT, and graduate school entrance exams such as the GRE, GMAT, LSAT, and MCAT.

Each of these tests has a component that tests for the ability to solve problems that are mathematical in nature. The test taker has to read, interpret, and understand situations that are described verbally and then answer a question related to the given information. Such problems are usually referred to as *word problems*. Sometimes the words describe specific arithmetic, algebraic, or geometric situations; other times the words require the test taker to find the mathematics inherent in the situation. These situations will require the test taker to apply logical reasoning, estimation skills, knowledge of measurement systems, arithmetic, algebra, and geometry. Many situations also require the interpretation of data, the use of statistics, and simple probability.

There are three formats for questions that are used on most standardized tests. They are *multiple choice, quantitative comparison*, and *short answer or free response*. To ensure accurate objective scoring for the large number of people taking the test, it is rare that long explanations of answers are required. Therefore, the best preparation for these tests not only includes a review of facts, concepts, and skills but also a review of strategies for answering questions in the three formats. This book offers the reader ways to improve his or her skills in both of these aspects of solving problems. Each chapter focuses on word problems that have a common element and format. Strategies for answering questions quickly are introduced as they apply to the example problems offered.

The first four chapters offer examples of problems in multiple-choice format. The fifth introduces questions that involve making a comparison between two given quantities. Within these chapters are problems involving estimation, measurement, the application of simple common formulas, algebra, ratios, proportions, geometry, probability, statistics, and logical reasoning. Each chapter focuses on a particular strategy that if used effectively, will help the reader understand and solve the problem faster than by using traditional methods. The index will help the reader locate problems of each mathematical area within the chapters.

Chapter 1 presents problems whose solutions can be found by working backward from the choices. Chapter 2 presents problems whose solutions can be found by substituting numbers into the problems. Chapter 3 presents problems whose solutions can be found by applying simple ratios and proportions. Chapter 4 presents problems whose solutions can be found by creating lists and looking for patterns. Chapter 5 presents problems from all areas in the two-column quantitative comparison format. Chapter 6 provides additional problem-solving practice through a collection of problems similar to those given in the preceding chapters. There are also several appendixes that will help the reader refresh basic arithmetic and algebra skills.

In most situations in this book, each example and additional problem has two solutions. The first is a "traditional" solution that solves the problem in a manner that would be appropriate if the problem were not in a multiple-choice format. These solutions are important to study as they will help the reader find solutions to free-response and/or short-answer questions. The second solution often involves the given choices and uses alternative strategies developed in each chapter.

In Chapter 4, the reader is presented with problems that are not all multiple-choice questions. The reader is instructed in how to fill in the grids for short answers that usually appear on standardized tests such as the PSAT and SAT.

Additional help with solving word problems can be found in *How to Solve Word Problems in Mathematics*. This book offers a traditional approach for a wide variety of problems encountered in mathematics.

Chapter 1

Making Use of Multiple Choices

Problems that appear on standardized tests such as the PSAT, SAT, ACT, and those created by state education departments are different from those that you have seen in math classes and in traditional textbooks or review books. Standardized test questions are different because they are designed to assess not only your knowledge of mathematics but also your ability to adapt quickly to unfamiliar situations. In fact, most of the problems on standardized tests are more verbal than computational, and they require you to know key vocabulary and concepts. Therefore, we will refer to any problem that verbally describes a situation as a *word problem* even if it doesn't quite look like the word problems you may have solved in math classes. It is important that you develop strategies to help you quickly understand verbal problem descriptions and arrive at the correct answer.

A critical aspect of a standardized test is that it is a timed test. That is, you have to complete the questions in the section in a prescribed time. In most tests, the average time for each question is a little more than 1 minute. For example, one section of the SAT requires answering 25 questions in 30 minutes, which allows an average of 1 minute and 12 seconds per question. Another part of the SAT requires answering 10 questions in 15 minutes, which allows an average of 1 minute and 30 seconds per question. Therefore, it is important to recognize situations in which you can use your intuition or make valid assumptions in order to solve the problem quickly.

In this chapter you will learn one way you can work through multiple-choice questions in which you are asked to select the answer from a list of given choices. Many times you can avoid doing tedious mathematics and, instead, find the answer by examining the choices given. Two key points to remember about multiple-choice questions are the following:

- The choices are part of the question. Use them to guide you to a better understanding of the problem.
- One of the choices is the correct answer to the problem. There are usually several ways to determine which one is correct without having to make tedious computations. In fact, sometimes, this recognition is really what is being tested.

In all word problems, look for key words or phrases in the problem statement. Usually, one word in the problem is the key to unlocking its mystery. The solutions to the examples will show you how to find the key idea. They will also show you how to apply traditional and alternative methods of using the choices themselves to arrive at the correct answer quickly.

Problems Involving Arithmetic and Properties of Numbers

Some problems will present simple situations that require familiar computation or concepts, but they may not contain enough actual information to use either directly. If you explore the numbers carefully, you may find that arithmetic is not at all necessary and the answer can be determined by recalling simple number facts.

Example I

If a and b are single digits, which of the following could be the quotient of $78ab6 \div 7$?

A 11,235   **B** 11,236   **C** 11,237   **D** 11,238   **E** 11,239

Solution I

Clearly, without knowing a and b, you cannot perform the division. However, if you look carefully at the choices, you will notice that the only difference among them is in their last digit. Remember, only one of them can be the correct answer. The answer multiplied by 7 must produce a number whose units digit is a 6. The only possible answer is choice D because $8 \times 7 = 56$ is the only product that produces a 6 in the units place. Therefore, the correct choice is D.

Example 2

If the area of a square room is represented by $18n$ where n is an integer, which of the following could be the length of the room?

A 5   **B** 6   **C** 7   **D** 8   **E** 9

Solution 2

The phrase "$18n$ where n is an integer" means that *the area is a multiple of 18*. The solution here is found by working backward with the choices. For each choice, the areas of the room would respectively be their squares: 25, 36, 49, 64, and 81. The only multiple of 18 in this list is 36. Therefore, the correct choice is B. (Did you notice that finding the value of n was not necessary?)

You will find that in many problems the numbers used are integers. Remember that the *integers* are the positive and negative whole numbers and zero: 0, ± 1, ± 2, ± 3, ± 4, and so on.

When facing multiple-choice questions, most people feel that they should answer the question first and then try to find the correct answer among the choices. However, on many standardized tests, the choices are not quite what you may expect. The following example demonstrates why it is important to study the choices before performing any calculations.

Example 3

If 3 is added to each of the digits of 65,021, then the resulting number is

A 3 more than 65,021.   **B** 15 more than 65,021.
C 30,000 more than 65,021.   **D** 33,333 more than 65,021.
E 98,354 more than 65,021.

Solution 3

The problem is not a difficult one and can be solved by creating the new number, which is 98,354. If you haven't really studied the choices, you might select choice E, which would be wrong. By looking at the choices first, you will notice that the problem is not about the computation. It is about understanding what occurs when you add 3 to each digit. Specifically, you are really adding the 5-digit number 33,333. The correct choice, D, is now obvious.

Problems Involving Algebra

Most of the time we think of word problems as always being solvable by creating algebraic equations to represent the situations. Solving the resulting equations is usually easy, but arriving at the equations can be difficult, especially if the verbal description of the problem confuses us. In each of the examples below, you will see the algebraic solution and a way to solve the problem without algebra by using the choices. An algebraic solution will also be the method used when the question requires a free response or short answer since there are no choices to use.

Example 4

The average of three numbers is 42. If one of the three numbers is 13, what is the sum of the other two?

A 29   **B** 71   **C** 85   **D** 97   **E** 113

Solution 4

The algebraic solution goes as follows: Let x and y be the other two numbers. The average is found by $(x + y + 13) \div 3 = 42$. Therefore, $x + y + 13 = 126$. $x + y$, which is what we are looking for, is $126 - 13 = 113$, or choice E.

Alternative Solution 4

You will not usually be asked to work with only one number if you are being asked about an average. The problem could cause immediate confusion if you start thinking about how you would find the other two numbers. If this is were the case or if you were unsure about how to proceed, you could try one of the choices and work backward to determine if it is correct.

In nearly all standardized tests, if the choices are all numbers, they usually appear in either ascending or descending order.

4

Therefore, it is a good strategy to start with the middle number, choice C. Using choice C in this problem, you would be assuming that the sum of the other two numbers is 85. (You could even suppose that the numbers were actually 40 and 45, although this is not necessary.) These numbers along with 13 would have a total sum of 98, and their average would be $98 \div 3 = 32\,^2\!/_3$. This is obviously incorrect, but you ought to realize that we have to increase the sum. It would not take you too much time to perform the same calculation with both choice D and choice E. As you develop more practice with using choices and working backward, you will develop an intuition about the remaining possible choices.

Example 5

If two positive numbers are such that the sum of their squares is 49 and the difference of their squares is 23, then the smaller number is

A $\sqrt{13}$.   **B** 6.   **C** 13.   **D** 26.   **E** 36.

Solution 5

The algebraic solution is as follows. Let x be the larger positive number, and let y be the smaller positive number. The conditions lead to the two equations $x^2 + y^2 = 49$ and $x^2 - y^2 = 23$. If you add the two equations together, you eliminate y^2 and have $2x^2 = 72$, or $x^2 = 36$. Therefore, $y^2 = 13$ and $y = \sqrt{13}$. Choice A.

Alternative Solution 5

By working backward with any of the choices, you could avoid the algebra. As in the previous example, suppose that the smaller integer was choice C, 13. Its square would be 169, and, since the sum of the squares is 49, it is an impossible situation. Therefore, you need a much smaller number. It makes sense to move immediately to the smallest number on the list, which is choice A, $\sqrt{13}$. Its square is 13. Therefore, the other square would have to be 36 in order for the sum to be 49. This also satisfies the condition that the difference of the squares is 23.

Example 6

A library has a collection of 315 biographies of politicians, actors, and sports stars. There are three times as many biographies of politicians as there are biographies of actors and five times as many biographies of actors as there are biographies of sports stars. How many biographies of sports stars are there?

A 105   **B** 35   **C** 21   **D** 15   **E** 5

Solution 6

The algebraic solution is as follows: Let x be the number of biographies of sports stars. Therefore, there are $5x$ biographies of actors and $3(5x) = 15x$ biographies of politicians. The total number of biographies is $x + 5x + 15x = 21x = 315$. x, our answer, is $315 \div 21 = 15$, choice D.

Alternative Solution 6

Working backward with one of the choices will avoid the algebra. Starting with choice C, suppose there are 21 biographies of sports stars. This implies that there are $5(21) = 105$ biographies of actors and $3(105) = 315$ biographies of politicians. Clearly, this is too many. Moving to a smaller choice, choice D will lead to the counts of 15, 75, and 225, which do indeed sum to 315.

Problems Involving Fractions and Percents

Fractions and percentages are often troublesome for test takers. The rules of making the computations in the correct fashion seem to get more confusing as time becomes short. Use of a calculator may help with the computation, but the calculator can't help the test taker determine which arithmetic operation to perform. However, the appendices in the back of this book can help you review and sharpen your computational skills with fractions, decimals, and percentages. In addition to practicing your skills with these, it is important to consider alternative ways to get to an answer quickly and correctly.

Example 7

A family's monthly budget includes $2,200 for housing and $500 for automobile expenses. What fraction of the total $6,000 budget remains for other expenses?

A $\dfrac{1}{12}$ **B** $\dfrac{11}{30}$ **C** $\dfrac{11}{20}$ **D** $\dfrac{22}{30}$ **E** $\dfrac{11}{12}$

Solution 7

There isn't anything difficult about solving this problem provided you understand that the amount spent on other expenses must be calculated and converted to its fractional part of $6,000. That is, the amount is $6,000 − ($2,000 + $500) = $3,300, and the fraction is $3,300/$6,000 = 11/20, choice C.

Alternative Solution 7

While the problem seems easy as you are taking your time reading this book, it may not feel that way under the pressure of a timed test. There are many ways to make a wrong calculation or misinterpret the problem. Each of the choices is an answer to a common mistake. By working backward through the choices, you can keep your focus on the problem. Starting with choice C, you suppose that the amount is 11/20 of $6,000 or $3,300. Keeping this number in front of you can help you focus on the amount for other expenses. This number added to $2,200 and $500 does indeed sum to $6,000. (Since there can be only one correct answer, there is no need to check any others.)

Example 8

Two years ago a small television sold for $180. Last year the price of this model was increased by 10%, and this year it was increased by 6%. Approximately what percent of the original price is the current price of this model?

A 108%   **B** 113%   **C** 116%   **D** 117%   **E** 160%

Solution 8

The first increase creates a price of $180 + 10\%$ of $180 = \$180 + \$18 = \$198$. The next increase is based on the new price and is $198 + 6\%$ of $198 = \$198 + \$11.88 = \$209.88$. The increase from the original price is $29.88, and the percentage increase is $29.88 \div 180 \times 100\% = 16.6\%$. The current price is 100% of the original price $+ 16.6\%$ or 116.6%. The correct choice is, therefore, D, or 117%.

Alternative Solution 8

By looking at the choices first, you should realize that the only possible choices are C and D. (Choices A and B are clearly too small, and choice E is way too large.) Comparing the remaining choices, you would have to ask yourself if the answer should be 16% or something larger. Since the percentages are compounded, the resulting percentage must be greater than the sum of the yearly increases. The only choice is D.

You might think that this alternative solution is fuzzy thinking. *On the contrary, it requires a deep understanding of the problem and an ability to think about situations quickly and effectively.* Remember, this is a large part of what standardized tests are trying to determine about your abilities.

Example 9

One-third of an estate was bequeathed to the family, two-fifths was bequeathed to a university, and the remaining $10,000 was left to a charity. How large was the original estate?

A $42,000   **B** $37,500   **C** $30,000   **D** $17,333   **E** $10,733

Solution 9

The algebraic solution is as follows: Let x be the amount of the estate. The family received $x/3$, and the university received $2x/5$. This represents

$$\frac{x}{3} + \frac{2x}{5} = \frac{5x}{15} + \frac{6x}{15} = \frac{11x}{15}$$

of the total amount. The remaining $10,000 would be $4x/15$ of the total amount. Therefore,

8

$$\frac{4x}{15} = \$10{,}000, \text{ and } x = \$10{,}000 \div \frac{4}{15} = \$37{,}500$$

or choice B.

Alternative Solution 9

Clearly there is plenty of room for error if your algebra and arithmetic of fractions aren't sharp. However, by working with the choices, you can once again avoid this. Starting with choice C, you suppose that the estate is $30,000. This leads to $10,000 going to the family and $12,000 to the university. The sum, $10,000 + $12,000 + $10,000, exceeds the $30,000 you started with. Therefore, you know you need one of the larger amounts. If you tried choice A, you would have the amounts $14,000 + $16,800 + $10,000, which is less than the $42,000 assumed. Even if you didn't have the time to check choice B, you could be confident that it is correct. (Of course, you should always check your answers when time allows!)

Problems Involving Geometry

A geometry problem involves the relationships between lines, angles, shapes, and solids. In most cases, diagrams will be given for you to use, and they are usually drawn to scale. In fact, in the directions of the test, you will find a phrase stating that *diagrams are drawn to scale unless told otherwise*. However, some geometric problems will describe a situation verbally, and you will have to visualize the situation yourself. In this regard, we can consider them to be word problems.

Example 10

Two angles measuring $p°$ and $q°$ are complementary. If $3p - 2q = 40°$, then the smaller angle measures

A $10°$.   **B** $40°$.   **C** $44°$.   **D** $46°$.   **E** $50°$.

Solution 10

Since the angles are complementary, we have $p + q = 90°$, and we can proceed with an algebraic solution. Multiplying this equation by 2, we have $2p + 2q = 180°$. Adding this to $3p - 2q = 40°$, we eliminate q and have $5p = 220°$, which gives us $p = 44°$. The complement of p is $46°$, which is larger than p. Therefore, $p = 44°$ is the smaller angle. Choice C.

Alternative Solution 10

By selecting any of the choices, you could determine the complement and substitute it into the given equation. Suppose, contrary to what you had been doing, you started with choice A, $10°$. Its complement would be $80°$. The given equation would be $3(80) - 2(10) = 220$, which is clearly not correct since this must be 40. Moving to choice B, $40°$, the equation becomes $3(40) - 2(50) = 20°$, or $3(50) - 2(40) = 70°$. Notice that you have to consider the equation in both ways since you don't know whether p is larger than q or vice versa. Moving to choice C, $44°$, gives the equations $3(46) - 2(44) = 50$ and $3(44) - 2(46) = 40$. Having found a pair that works, you need not check the remaining choices. (There can be only one answer!)

Example 11

The average of the degree measures of the two angles of an isosceles triangle is $54°$. Which of the following could be the measure of the third angle?

A $36°$ **B** $72°$ **C** $108°$ **D** $126°$ **E** $144°$

Solution 11

The key fact about an *isosceles triangle* is that two sides are equal in length and the angles opposite these sides are equal in measure. (These angles are called the *base angles*, and the third angle is referred to as the *vertex angle*.) The algebraic approach to the problem is as follows: Let the three angles of the triangle have measures $x°$, $x°$, and $y°$. Therefore, we have the equation $2x + y = 180°$. If two of the angles have an average of $54°$, then their sum is $108°$. (That is, the sum divided by 2 must be $54°$.)

There are two possibilities, either

$$2x = 108 \qquad \text{or} \qquad x + y = 108$$

In the first case, each of the angles used in the average is $x = 54°$, and the third angle would be $y = 72°$. In the second case, by subtracting the equations, we have the base angle $x = 72°$ and the vertex angle $y = 36°$. The third angle, not used in the average, would be the other $72°$ base angle. Therefore, we have two possible isosceles triangles that could solve the problem, one with the angles $54°$, $54°$, and $72°$ and the other with the angles $72°$, $72°$, and $36°$. However, in both cases, the third angle is $72°$, choice B.

Alternative Solution 11

By using the choices, we can gain some immediate intuition about the situation. With choice C, we are supposing that the third angle is $108°$. Since this is an obtuse angle, it must be the vertex angle of the isosceles triangle. The two congruent base angles must each be $36°$. Among the three angles, $36°$, $36°$, and $108°$, there is no pair that has an average of $54°$. It is reasonable that you would want numbers closer to $54°$; therefore, the next logical choice is B, $72°$. You could assume that this is the vertex angle, and you would have the isosceles triangle with the angles $54°$, $54°$, and $72°$. Since the other two angles do indeed have an average of $54°$, you have found the correct choice. (If you had assumed that $72°$ was a base angle, you would have the other isosceles triangle with the measures $36°$, $72°$ and $72°$ and, once again, the other two angles have an average of $54°$.)

An important aspect of problem 11 is that there were two different possibilities for the triangle described in the problem. You might be inclined to call a problem like this a "trick question." However, recognizing *when to consider all possibilities* is an important problem-solving skill in all studies and careers. This is another ability that is being measured by the standardized test. In a later chapter, we will discuss when you can make valid assumptions that eliminate some possibilities. Remember to always ask yourself, *"Is there another way to satisfy the situation?"*

For each of the following additional problems, you will find a traditional solution using appropriate formulas, arithmetic, and algebra. You will also have an alternative solution showing how to avoid tedious computations or algebra. You should study both solutions carefully. *The traditional solutions will be necessary for solving problems that are not in the multiple-choice format and thus require you to produce an answer of your own design.* The alternative solutions will offer more hints and strategies for helping you use the choices to recall important mathematical ideas necessary for solving the problem. Try each problem yourself before looking at the solutions.

Additional Problems

1. There are red and black cut-out shapes in a bag. If $2/3$ of the shapes are black, and $1/2$ of the red shapes are triangles while the remaining 8 are squares, how many objects are in the bag?
   A 32   **B** 48   **C** 54   **D** 64   **E** 80

2. Joe, Rob, and Allison have been collecting empty soda bottles. Joe and Rob have 30 together, Allison and Rob have 42 together, and Joe and Allison have 16 together. What is the least amount that any of the three have?
   A 1   **B** 2   **C** 10   **D** 12   **E** 14

3. A shopkeeper offers a 25% discount on the marked price on an item. In order for the item to now cost $48, what should the marked price be?
   A $12   **B** $36   **C** $60   **D** $64   **E** $75

4. If the circulation of a local newspaper was 2,500 last year and is now 4,000, what is the percent change of the circulation?
   A 15%   **B** 60%   **C** 62.5%   **D** 150%   **E** 600%

5. The difference of two positive integers is 9, and the difference of their squares is 189. What is the sum of the numbers?
   A 198   **B** 180   **C** 108   **D** 30   **E** 21

6. A hiker spent 6 hours following a trail to its end and returning to the starting point along the same path. If she walked three times as fast on the return trip, for how long did she walk from the starting point to the end of the trail?
   A 3 hours   **B** 3.5 hours   **C** 4 hours   **D** 4.5 hours   **E** 5 hours

7. A bus takes 3 hours to go from Clarkson to Jamesberg, $2\frac{1}{4}$ hours to go from Jamesberg to Evansville, and the remainder of the time to go from Evansville to Folger Springs. If the average speed of the bus was 52 miles per hour for the entire 468-mile trip from

Clarkson to Folger Springs, how much time was spent traveling from Evansville to Folger Springs?

A 2 hours   **B** 3 hours   **C** $3\frac{3}{4}$ hours   **D** $4\frac{3}{4}$ hours   **E** 9 hours

8. A rectangular fishtank holds a maximum of 27 cubic feet of water. If the tank is 8 feet long and 15 inches wide, how many feet deep must the water be if the tank is $^2/_3$ full?

   A 2/3   **B** 1   **C** 1.25   **D** 1.8   **E** 2.7

9. In a triangle, the square of one side is equal to the difference of the squares of the other two sides. The largest angle of this triangle measures

   A 30°.   **B** 45°.   **C** 60°.   **D** 75°.   **E** 90°.

10. Points $(6, -2)$ and $(a, 6)$ are on a line with a slope of $^4/_3$. What is the value of a?

   A −2   **B** 4.5   **C** 9   **D** 12   **E** 15

11. Six congruent squares are arranged to form a single rectangle whose perimeter is 168 inches. Which of the following is possible for the number of square inches in the area of one of these squares?

   A 12   **B** 14   **C** 144   **D** 168   **E** 196

12. From the following choices, which represents the greatest possible distance between two points on a circular table top with a circumference of approximately 12.6 feet?

   A 2 feet   **B** 3.14 feet   **C** 3.9 feet   **D** 4π feet   **E** 16 feet

13. If a circle has a circumference of 6π, and a triangle has the same area, what is the length of the altitude of the triangle drawn to a side whose length is 8?

   A π **B** 2π **C** 2.25π **D** 2   **E** 2.25

14. When the 100 members of a set of numbers are multiplied together, the product is negative. What is the number of negative integers in the set?

   A 0   **B** 2   **C** 99   **D** 100

   E The answer cannot be determined from the given information.

15. Which of the following numbers is not divisible by a square of a prime number?

   A 420   **B** 630   **C** 770   **D** 1,260   **E** 1,470

Solutions to Additional Problems

Solution to Problem 1

Trying to set up an algebraic model for this problem will get confusing. Let x be the number of objects in the bag. The number of black objects is $^2/_3x$, which implies that $^1/_3x$ is the number of red objects. The number of triangles is $\frac{1}{2}$ of $\frac{1}{3}x = \frac{1}{6}x$, which is 8, since the remaining squares are also $\frac{1}{2}$ of the red objects. If $\frac{1}{6}x = 8$, then $x = 48$, choice B.

Alternative Solution to Problem 1

Using the choices will be more direct. With choice C, we are supposing that there are 54 objects in the bag. The number of black objects is $\frac{2}{3}$ of $54 = 36$. Therefore, there are 18 red objects, and 9 would be triangles. There are 9 remaining, which contradicts the problem. It is reasonable to go to a smaller choice since we want a smaller number remaining. Working this through with choice B, the number of black objects is $\frac{2}{3}$ of $48 = 32$, leaving 16 reds. The number of triangles would be 8, and the remaining 8 would be squares. The conditions of the problem are satisfied, and we have our correct answer.

Solution to Problem 2

Using algebra, we let J be the amount that Joe has, R be the amount that Rob has, and A be the amount Allison has. The information in the problem gives the three equations

$$J + R = 30 \qquad A + R = 42 \qquad J + A = 16$$

If we add all three equations together, we get $2(J + A + R) = 88$, or $J + A + R = 44$. Since the third equation tells us that $J + A = 16$, we have $R = 28$. Therefore, $J = 2$ from the first equation, and $A = 14$ from the second. The least amount had by any of the three is 2, choice B.

Clearly, the algebra used in the solution is not traditional, and you may have never solved a system of three equations by adding equations where variables aren't eliminated. You can work with two equations and eliminate a variable. That is, by subtracting the second equation from the first, you have $J - A = 12$. Adding this equation to the third gives us $2J = 28$, or $J = 14$, and you can find the values for A and R from here.

Alternative Solution to Problem 2

Working backward from the choices will eliminate the algebra. Since we are looking for the least amount, it makes sense to start with the smallest of the choices, 1. The information tells us that the combinations with Joe have the smaller amounts. Therefore, it is logical to assume that Joe has the least amount. Assume Joe has 1, Allison has 15, and Rob has 29. This, however, leads to Allison and Rob together having 44, which contradicts the information in the problem. Moving to the next smallest choice, B, Joe would have 2 leading to Allison having 14 and Rob having 28. This satisfies the condition that Rob and Allison together have 42, and we have found our answer.

14

Solution to Problem 3

A 25% discount means that the item will sell for 75% of the marked price. Let x be the marked price, which generates the equation $0.75x = 48$, or $x = 48 \div 0.75 = \$64$, choice D.

Alternative Solution to Problem 3

By using the choices as the possible marked prices, you will be directly computing the discounted price. This will avoid the potential errors involved with the computation. Trying choice C, we suppose that the marked price is $60. The discount would be 25% of $60 = $15, and the new price would be $45. This is too small, but close. It is reasonable to move to a higher marked price, and choice D, $64, will satisfy the problem.

Solution to Problem 4

The *percent change of a quantity* is the percentage that the difference of the amounts is based on the original amount. The change here is 1,500, which is $1,500/2,500 \times 100\% = 60\%$, choice B.

Alternative Solution to Problem 4

As you look at the choices before starting the problem, you should be wondering whether the percent change could be more than 100%. A key fact to remember is that a *100% change in a quantity means that the quantity doubled.* That is, you have added 100% of the original amount to itself to get the new amount. Since the amount did not double here, you can eliminate choices D and E. This also will help to remind you that you are concerned with the amount by which the original increased. If you try choice C, you would compute 62.5% of $2,500 = 1,562.5$, which is too much, but close. Choice B becomes the obvious answer. Checking involves computing 60% of $2,500 = 1,500$, the actual amount of increase.

Solution to Problem 5

Algebraically we let x be the larger integer and y be the smaller integer. The information tells us that $x - y = 9$ and $x^2 - y^2 = 189$. The key to the algebra is to remember that $x^2 - y^2 = (x - y)(x + y)$. This gives us $189 = 9(x + y)$, and, therefore, $x + y = 21$, choice E.

Alternative Solution to Problem 5

Using a choice for the sum of the numbers, we would attempt to find a pair of positive integers that have that sum and a difference of 9. We would have picked the right choice if the difference of the squares is indeed 189. To avoid too much computation, it makes sense to start with the smallest number, 21. The pair having a sum of 21 and a difference of 9 is 15 and 6. The difference of the squares is $225 - 36 = 189$, and we are done. (Note that the other choices are all even. Working with the sole odd choice is reasonable. In fact, none of the other choices can be broken down into a pair of integers with that sum and a difference of 9.)

Solution to Problem 6

An algebraic solution involves the formula distance = rate × time. There are two rates and two times in the problem while the distances going and returning are the same. Let T_1 and R_1 be the time and distance going and T_2 and R_2 be the time and distance returning. We have the equation $T_1 R_1 = T_2 R_2$. We also have that $R_2 = 3R_1$. Using this in the first equation, we have $T_1 R_1 = T_2(3R_1)$. Dividing both sides by R_1 gives us $T_1 = 3T_2$. The problem tells us that $T_1 + T_2 = 6$. We now have $3T_2 + T_2 = 6$, or $4T_2 = 6$ or $T_2 = 6/4 = 1.5$. Therefore, $T_1 = 6 - 1.5 = 4.5$, choice D.

Alternative Solution to Problem 6

Using a choice and working backward still requires knowing that distance = rate × time and that the distance going is the same as the distance returning. There is also a need to use R and $3R$ as the respective rates. Assuming choice C, the time taken was 4 hours, and the first leg of the journey covered a distance of $4R$ while the return trip covered a distance of $2(3R) = 6R$. This cannot be correct since *the distances have to be equal*. You should realize that, in order to bring these numbers closer, the time on the first leg of the journey has to be greater. Using choice D, 4.5 hours, we have that the distance for the first leg is $4.5R$ and the distance for the second leg is $1.5(3R) = 4.5R$. The distances are indeed the same, and we have solved the problem.

Solution to Problem 7

Since the average speed of the trip was 52 miles per hour and the distance was 468 miles, using the formula distance = rate × time, we have $468 = 52 \times$ time. Therefore, the total time for the trip was 9 hours. The time from Clarkson to Evansville is $9 - (3 + 2^1/_4) = 9 - 5^1/_4 = 3^3/_4$, choice C.

Alternative Solution to Problem 7

You might be inclined to start with choice B, 3 hours, since this appears at first to be easy to work with. Your total time would be $8^1/_4$ hours. The average speed would be 468 miles ÷ 8.25 hours = 56.7272..., which is too large. Moving to choice C, $3^3/_4$ hours, gives us a total time of 9 hours and 468 ÷ 9 = 52, which satisfies the condition of the problem.

16

Solution to Problem 8

It is necessary to recognize that *holding water* and *cubic feet* both imply that we need the volume of the tank. The formula for the volume of a rectangular solid is volume = length × width × height. It is also important to make sure that all the measurements are in the same units—in this case, feet. The width of the tank is 15 inches, which is equivalent to $15 \div 12 = 1.25$ feet. Since we want only $2/3$ of the maximum capacity, we are seeking a volume of $2/3 \times 27 = 18$ cubic feet. Using the formula and letting h be the height of the water, we have $18 = 8 \times 1.25 \times h$ or $18 = 10h$ and $h = {}^{18}/_{10} = 1.8$ feet, choice D.

Alternative Solution to Problem 8

By looking at the choices first and realizing that they are small numbers, you may be reminded of the need to convert 15 inches to 1.25 feet. If you do not convert the width from inches to feet, none of the choices would give you the correct answer. That is, $8 \times 15 = 120$, and computing the volume with any of the choices as the height produces numbers greater than 27!

When none of the choices work, you must reread the problem carefully and look for items that you failed to consider the first time.

Once you've made the conversion to 1.25 feet, you can compute the volume for any of the choices. Note that choice E produces a volume of 27. If you selected this, you would have failed to utilize the fact that the water fills only $2/3$ of the tank. Choice D gives 18 cubic feet, and the check would be that $2/3$ of 27 = 18.

Solution to Problem 9

The first side mentioned could not be the largest side of the triangle since it is found by subtracting the squares of the other two. If we let this side be x and the other two y and z with z being the largest, we have $x^2 = z^2 - y^2$, which gives us $x^2 + y^2 = z^2$. This is only the case with a right triangle, and, therefore, the triangle must have a $90°$ angle and the other angles must be acute. The correct answer is choice E.

Alternative Solution to Problem 9

Since we are looking for the largest among the choices, it makes sense to start with the largest choice and work down. This prompts the question "Must the triangle be a right triangle?" Since the square of one side is the difference of the squares of the other two sides, then the square of the largest side must be the sum of the squares of the other two sides, which is the Pythagorean theorem satisfied only by a right triangle.

Solution to Problem 10

The slope of a line is the fraction formed by the difference in the y coordinates over the difference in the x coordinates. This gives us $(6 - {}^-2)/(a - 6) = {}^4/_3$ or $8/(a - 6) = {}^4/_3$. Cross multiplying, we have the equation $4a - 24 = 24$, leading to $4a = 48$ and $a = 12$, choice D.

Alternative Solution to Problem 10

If you have an understanding about slope, you should realize that for the slope to be a positive number, which is the case here, an increase in the y coordinate requires an increase in the x coordinate. This rules out choices A and B. Using choice C, $a = 9$, we can compute the slope to be ${}^8/_3$. This is not correct, so we try choice D, $a = 12$, and the slope is ${}^8/_6$, which is the ${}^4/_3$ we seek.

Solution to Problem 11

The key word in the question is *possible* reminding us to *consider all possibilities*. There are two possible arrangements, a single row of the six squares or two rows of three. Let x be the measure of the side of a square. The first arrangement would have a perimeter of $14x$, and the second arrangement would have a perimeter of $10x$. In the first case, $14x = 168$ and $x = 12$. The area of a square would be 144 square inches, which is choice C. In the second case, we have $10x = 168$ and $x = 16.8$. The area of the square would be larger than $16^2 = 256$, which is not among the choices. The correct answer is choice C.

Alternative Solution to Problem 11

Using choice C, 144 square inches, implies that each square has a side of 12 inches. You still have to consider both possible arrangements as described above. The first arrangement has a perimeter of $12 + 12 + 72 + 72 = 168$, which is exactly what is given. You could stop your work now or continue to explore the other possibility. *Remember, there can be only one correct answer to the problem, and, if you've found a choice that solves the problem, it must be the answer. Since time is important, the strategy is to take this answer and move on to the next question.*

Solution to Problem 12

The greatest distance between two points in or on a circle occurs where the points are the endpoints of a diameter. The circumference of a circle is equal to $\pi \times$ diameter. Therefore, the diameter in this problem is $12.6 \div 3.14$, which is approximately 4. The key phrase in the problem is *from the following choices*, and choice C, 3.9 feet, represents the largest possible distance.

Alternative Solution to Problem 12

The choices are full of hints to remind you about measurements concerning a circle. For example, choice B is the numerical equivalent of π, and choice D is approximately 12.6 to remind you of the formula given above. You should also realize that the straight-line distance between any two points in or on the circle has to be less than the circumference of the circle, which will eliminate choices D and E immediately. It would be logical to start with choice C since we are looking for the greatest distance. Using 3.9 as a possible diameter, we find that this gives us a circumference of approximately $3.9 \times 3.14 = 12.3$ feet. Any of the other two choices would give us a smaller circumference. Therefore, choice C must be the correct answer.

Solution to Problem 13

From the formula for the circumference of a circle, circumference = $\pi \times$ diameter, we have that the diameter is 6 and the radius is 3. The area of a circle is πr^2 and, in this case, 9π. The area of a triangle is $\frac{1}{2} \times$ base $\times$ height. Using h as the length of the altitude, we now have the equation $\frac{1}{2}(8)h = 9\pi$ or $4h = 9\pi$. Then $h = 9\pi \div 4 = 2.25\pi$, choice C.

Alternative Solution to Problem 13

Realizing that the area of a circle involves π, we should try A, B, or C before D or E. The area of the triangle can be computed using each of these choices and the base of 8. That is, for choice A the area is $\frac{1}{2} \times 8 \times \pi = 4\pi$, for choice B the area is $\frac{1}{2} \times 8 \times 2\pi = 8\pi$, and for choice C the area is $\frac{1}{2} \times 8 \times 2.25\pi = 9\pi$.

For each of these, you can determine the radius of the circle. That is, for choice A, $r^2 = 4$ and $r = 2$. The circumference would have to be $C = 2\pi r = 2\pi \times 2 = 4\pi$, which is not what is given. For choice B, $r^2 = 8$ and $r = \sqrt{8}$, making the circumference $2\sqrt{8}\pi$, which, again, is not what is given. For choice C, $r^2 = 9$ and $r = 3$, making the circumference $2\pi \times 3 = 6\pi$, which corresponds to the given.

Solution to Problem 14

Since the product of two negative numbers is positive, then the product of any even amount of negative numbers will also be positive. Therefore, you need an odd amount of negative numbers. Choice C seems reasonable. However, the correct answer to this problem is E! The important word in the question is *integer*. The definition of the set does not specify that the members are integers.

Alternative Solution to Problem 14

If you approach this problem by examining each of the choices, the question you have to ask is, "Is it possible for the set to have this many integers?" When you have a choice such as E, you have to consider the possibility that there is something deeper in the problem before you jump to an answer. Starting with choice A, you would ask yourself, "Can the set have no integers?" Since the set does have 100 numbers, you should be prompted to think of the definition of an integer (a negative or positive number that is not a fraction or decimal). This is clearly possible as is any of the other choices since producing a negative product does not require avoiding fractions or decimals. If more than one answer satisfies the conditions of the problem, then E becomes the correct answer. (Remember this when you reach Chapter 6.)

Solution to Problem 15

You have no choice but to examine each of the choices to find the correct answer. Every integer can be factored into a product of prime numbers. Each of the numbers in the list is divisible by 10, and, therefore, each of their factored forms includes 2×5. After dividing by 10, the choices become 42, 63, 77, 126, and 147. If any of these have another factor of 2, then the original number was divisible by 4 and can be eliminated. Clearly, 77 is factorable only as 7×11 and the original number as $2 \times 5 \times 7 \times 11$ having no squares. Thus, choice C, 770, is the correct answer.

Alternative Solution to Problem 15

You can list the squares of the first several primes 4, 9 , 25, 49, 121, and 169, and divide each choice by these. For choice A, 420 is divisible by 4. For choice B, 630 is divisible by 9. For choice C, none from the list divide the number evenly. Before continuing, you should move to choice D, 1,260, which is divisible by 4 or by 9. For choice E, 1,470 is divisible by 49. Therefore, the correct answer is C.

Chapter 2

Working with Many Variables

Many multiple-choice word problems on standardized tests do not involve specific numbers. The statement of the problem contains information about variables, and the choices are all expressions containing those variables. It may appear that the question requires you to use algebra on the variables to arrive at the correct expression. However, the test may be actually assessing your understanding that variables represent numbers. *What is true in a variable question should be true for all numbers.* Therefore, it is okay to substitute numbers for the variables in the statement of the problem and work the problem with these numbers to find the expression among the choices that produces the same result. Let's call this strategy *substitution*.

The first example involves only a basic understanding of arithmetic so as to demonstrate the preceding idea, and the correct choice is easily found by simply performing the described operations.

Example I

Let x and y be two numbers whose sum is not zero. If x is multiplied by y and the product is then divided by the sum of the two numbers, what is the result?

A $\dfrac{x+y}{xy}$ **B** $\dfrac{xy}{x+y}$ **C** $xy + \dfrac{x}{y}$ **D** $(x+y)xy$ **E** $\dfrac{x}{y}(xy)$

Solution I

Since the answer must be the expression that holds true for specific values of x and y, we can substitute for these variables with numbers. Note that not all numbers will be usable since the computation contains a division and we cannot divide by zero. Therefore, we will not use zero in this problem.

Let $x = 2$ and $y = 3$. The problem is now restated, "If 2 is added to 3 and the product, 6, is then divided by the sum of the two numbers, 5, what is the result?" Clearly, the calculation is $6/5$. When subbing in 2 and 3 into the choices, they become the following:

A $\dfrac{5}{6}$ **B** $\dfrac{6}{5}$ **C** $6\dfrac{2}{3}$ **D** 30   **E** 4

The only matching choice is B, and, therefore this must be the correct choice. The process would be the same no matter what numbers you used, other than 0. (Try it.)

In the examples that follow, you will see a traditional solution followed by one that uses the strategy discussed above. It is rare that these types of questions would require free-response or short-answer solutions. However, just in case, you should become adept at both types of solutions.

Problems Involving Fractions and Percents

Example 2

Before the latest delivery of home heating oil, the homeowner's oil tank was $1/3$ full. If x gallons were delivered to fill the tank, how many gallons of oil, in terms of x, does the tank hold?

A $3x$ **B** $\dfrac{2x}{3}$ **C** $\dfrac{3x}{2}$ **D** $\dfrac{x+2}{3}$ **E** $3x - \dfrac{x}{3}$

Solution 2

The formal algebraic solution is as follows: Let T be the amount that the tank holds when full. The homeowner already had $T/3$ gallons before delivery. When the x gallons were delivered, the tank was full, and it had $T/3 + x$ gallons. Therefore, the equation is

$$T = \frac{T}{3} + x$$

Subtracting $T/3$ from both sides gives us $2T/3 = x$. Multiplying both sides by $^3/_2$ solves the problem for us—namely,

$$T = \frac{3x}{2}$$

or choice C.

You probably realize at this point how confusing this can become when working with the variables. Consider the substitution approach.

Alternative Solution 2

In this problem we need to substitute for the capacity of the tank. Since we are starting with $^1/_3$ of the capacity, we ought to use a number divisible by 3, say, 30. Therefore, the situation is that the tank had $^1/_3$ of $30 = 10$ gallons before delivery, and it took $x = 20$ gallons to fill. When x is 20, the choices become the following:

A 60   **B** $\dfrac{40}{3}$ **C** 30   **D** $\dfrac{22}{3}$ **E** 51 $\dfrac{1}{3}$

Therefore, the correct choice is C. Do you agree that this is somewhat simpler?

Example 3

If p is 20% of q, and q is 25% of r, and r is 80% of s, which of the following is the ratio of p to s?

A 1:4   **B** 1:5   **C** 1:20   **D** 1:25   **E** 1:50

Solution 3

The formal solution is as follows:

$r = 80\%$ of $s = 0.8s$

$\rightarrow q = 25\%$ of $r = 0.25r = 0.25 \times 0.8s = 0.2s$

$\rightarrow p = 20\%$ of $q = 0.2q = 0.2 \times 0.2s = 0.04s$

Therefore, the ratio $p{:}s = 0.04s/s = 0.04 = {}^1/_{25}$.

There are many places to make an error in this process, and, therefore, substitution may well be preferred.

Alternative Solution 3

Since the numbers are variable and the ratios are specific numbers, we can expect that our answers would be the same for *any* numbers we are working with. The key to this problem is to work backward from a value for s. *Since we are working with percentages, the best substitution to start with is $s = 100$.* Therefore, $r = 80\%$ of $100 = 80$, q is 25% or $\frac{1}{4}$ of $80 = 20$, and p is 20% or $\frac{1}{5}$ of $20 = 4$. Therefore, the ratio $p : s = 4 : 100 = 1 : 25$, choice D.

Example 4

A bus usually travels a route of m miles in h hours and 10 minutes. On the return trip along the same route, the bus makes one additional stop that adds 5 minutes to the trip. In terms of m and h, what is the average speed of the bus during the entire route in miles per hour?

$$\textbf{A} \ \frac{m}{h} + 25 \qquad \textbf{B} \ \frac{m}{h + 25} \qquad \textbf{C} \ \frac{2m}{2h + \dfrac{1}{10}}$$

$$\textbf{D} \ \frac{2m}{2h + \dfrac{5}{12}} \qquad \textbf{E} \ \frac{2m}{2h + 25}$$

Solution 4

The total mileage for the trip is $2m$ miles. The total time for the trip is $2h$ hours and 25 minutes. The minutes have to be converted to hours. Thus,

$$25 \text{ minutes} = \frac{25}{60} = \frac{5}{12} \text{ hours}$$

Therefore, the round trip took

$$2h + \frac{5}{12} \text{ hours}$$

The average speed is

$$\frac{2m}{2h + \dfrac{5}{12}}$$

or choice D.

Alternative Solution 4

Since we will be dividing into m, it is a good idea to pick a value with many factors, say, 120. We can let $h = 3$. Therefore, the round trip was 240 miles, and it took 6 hours and 25 minutes to complete. Working with actual numbers may remind you to convert the minutes to hours. If you use a calculator to make the conversion, you have approximately 6.42 hours. The average speed is approximately $240/6.42 = 37.38$ miles per hour.

Substituting $m = 120$ and $h = 3$ into the choices, we have the following approximate values:

A 65   **B** 4.29   **C** 39.34   **D** 37.40   **E** 7.74

and choice D is the closest approximation.

You may be uncomfortable with an approximation as a basis for an answer. However, this concern should not arise if only one choice is near enough to the calculation, which is the case in the preceding problem. You can try the problem again with different values for m and h and determine that choice D will again be the closest approximation.

The following example uses a ratio as its basis. The key fact about ratios is that *a ratio is a fraction. Thus, "x is to y" and "x : y" are equivalent to x/y*. (In Chapter 3 you will discover strategies to help with a variety of problems involving ratios and proportions.)

The next example also introduces you to a problem that is most likely very different from problems you've seen in classrooms. In these types of problems, you are given three statements identified by roman numerals. (For easy reference, let's agree to call these *roman numeral questions*.) The choices A through E state that specific combinations of these statements answer the question in the problem. When you approach these problems, you must remember to *consider all possibilities*.

Example 5

If the ratio $a{:}b$ is greater than 1, which of the following is true?

I. $a > b$.   II. $\dfrac{a^2}{b^2}$ is greater than 1.   III. ab is positive.

A I only   **B** II only   **C** III only   **D** II and III only

E I, II, and III

Solution 5

The ratio $a{:}b$ is the fraction a/b. If a and b are both positive numbers, then all three statements are true. However, it is possible for a and b to both be negative and have a ratio that is greater than 1. In this case, statement I is false while the others remain true. Therefore, choice D is the correct answer.

Alternative Solution 5

Since the problem says so little about a and b individually, you should think of this as a prompt to substitute different numbers for them to satisfy the condition. You should always be thinking about using *positive integers, negative integers, zero, and fractions* as values. First trying positive integers such as $a = 4$ and $b = 3$ makes the statements become the following:

I. $4 > 3$.   II. $\dfrac{16}{9}$ is greater than 1.   III. 12 is positive.

Using negative integers $a = -4$ and $b = -3$ makes the statements become the following:

I. $-4 > -3$.   II. $\dfrac{+16}{+9}$ is greater than 1.   III. $+12$ is positive.

Clearly, statement I is not always true. Using zero would not satisfy the conditions of the problem, and using fractions won't alter the nature of the statements.

Problems Involving Algebra

Example 6

If x and y are any two nonzero real numbers, what is $x\%$ of y divided by $y\%$ of x?

A xy **B** $\dfrac{x}{y} \times 100$ **C** $\dfrac{xy}{100}$ **D** $10{,}000xy$ **E** 1

Note that the problem states that x and y are *real numbers.* Students who have had advanced math courses may remember that there is a larger set of numbers than those we ordinarily work with called the *complex numbers*, which includes imaginary numbers involving the number $i = \sqrt{-1}$. Most standardized tests that test the general population won't involve complex numbers, and you therefore should assume that the numbers you are working with are real numbers. A *real number* is any whole number, integer, fraction, or infinitely nonrepeating decimal.

It is important for the author of the problem to tell you that neither x nor y can be zero since the problem involves division. We know that division by zero is an impossibility, and, if asked to do so, there is an undefined result.

Solution 6

Algebraically,

$$x\% \text{ of } y \text{ is } \frac{x}{100} \times y \text{ and } y\% \text{ of } x \text{ is } \frac{y}{100} \times x$$

The division is

$$\frac{xy}{100} \div \frac{yx}{100} = \frac{xy}{100} \times \frac{100}{yx}$$

which reduces to 1 in all cases.

Alternative Solution 6

Substitution makes this problem very easy as follows. Since we are dealing with percentages, let us choose numbers that make the calculations easy. Suppose $x = 50$ and $y = 100$. The problem now becomes "What is 50% of 100 divided by 100% of 50?" The arithmetic is simply 50 divided by 50 which is 1. When substituting these values into all of the choices involving x and y, we find that they become the following:

A 5,000   **B** 50   **C** 50   **D** 50,000,000   **E** 1

which doesn't change for any values of x and y.

Clearly, choice E must be the answer since the problem has to be satisfied for any values of x and y.

27

Example 7

The nonzero numbers a and b are such that $a = b^4$. If the value of b is tripled, the new value is which of the following?

A a raised to the fourth power **B** a multiplied by 3
C a multiplied by 4 **D** a multiplied by 12
E a multiplied by 81

Solution 7

Algebraically, we let the new value be k, and we have $k = (3b)^4 = 3^4 b^4 = 81a$, choice E.

(Note again how the fact that a and b are nonzero is important. If they were zero, then the new value would also be zero.)

Alternative Solution 7

Using substitution, we can let $b = 2$ (avoid using 1 as a substitution as it has unique properties in arithmetic). Thus, $a = 2^4 = 16$. If b is tripled, we have a new value of $6^4 = 1,296$. Comparing this to the original value, we find that $1,296/16 = 81$, choice E.

Example 8

A television that costs d dollars can be paid for in either 8 or 12 equal payments. How many dollars less would each payment be if a buyer chose to make 12 equal payments compared to 8?

A $\dfrac{d}{96}$ **B** $\dfrac{d}{24}$ **C** $\dfrac{d}{12}$ **D** $\dfrac{d}{8}$ **E** $\dfrac{d}{4}$

Solution 8

Algebraically, each of the 12 payments is $d/12$, and each of the 8 payments is $d/8$. The subtraction is

$$\frac{d}{8} - \frac{d}{12} = \frac{12d}{96} - \frac{8d}{96} = \frac{4d}{96} = \frac{d}{24}$$

or choice B.

Substitute a number for d that is divisible by 8 and by 12, say, 48. (This may not be a realistic price for a television, but such accuracy is not essential in solving the problem.) Therefore, each of the 12 payments is $4, and each of the 8 payments is $6, and the difference is $2. Using $d = 48$, the choices become the following:

A $\dfrac{48}{96} = \dfrac{1}{2}$ **B** $\dfrac{48}{24} = 2$ **C** $\dfrac{48}{12} = 4$ **D** $\dfrac{48}{8} = 6$ **E** $\dfrac{48}{4} = 12$

Choice B must be the answer.

Problems Involving Geometry

Once again, we will refer to some problems as "word problems" simply because they are described verbally. The following problem may not require translating verbal sentences into algebraic expressions, but algebraic expressions do indeed need to be created from geometric relationships in order to solve the problem directly. You will also see in the alternative solution that substition will help you avoid the pitfalls of using algebra.

Example 9

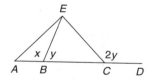

In the figure above, segment *EB* creates angles measuring $x°$ and $y°$ on segment *ABCD*. If the measure of angle *ECD* is $2y°$, what is the measure of angle *ECB* in terms of x and y?

A $2x$ **B** $180 - y$ **C** $x + y$ **D** $x - y$ **E** $y - x$

Whether you use substitution or work with the variables as given, you still have to know some basic facts about angles in and around a triangle. In this problem, key facts are the following:

- *An exterior angle formed by extending one side of a triangle is equal to the sum of the measures of the two angles within the triangle not adjacent to it.*
- *Two adjacent angles lying on the same straight line are supplementary (the sum of their measures is 180°).*

Solution 9

To start, let the measure of angle ECB be z. The goal is to solve for z in terms of x and y. Using the variables and the key fact above concerning exterior angles, the measure of $\angle BEC$ must also be $y°$ since $m\angle ECD = m\angle EBC + m\angle BEC$ or $2y = y + m\angle BEC$. $\angle ABE$ is also an exterior angle of triangle BEC and is equal to $m\angle BEC + m\angle EBC$. That is, $x = y + z$. Therefore, $z = x - y$, choice D.

Alternative Solution 9

Substitute values that satisfy the fact that x and y form a 180° angle. Suppose $x = 100$ and $y = 80$. Therefore, the measure of angle ECD is 160°, and $m\angle BCE$ must be 20°. Using these values for x and y in the choices, the choices become

A 200   **B** 100   **C** 180   **D** 20   **E** −20

It is clear that the correct choice is D.

Example 10

In the figure above, the three circles are congruent. If the diameter of the circle is d centimeters, how many square units are in the area of the shaded region in terms of d?

A $d(6 - 2\pi)$ **B** $2d^2 - 2\pi$ **C** $d^2\left(2 - \dfrac{\pi}{2}\right)$ **D** $2d^2(1 - \pi)$

E $2d^2\left(2 - \dfrac{3\pi}{2}\right)$

In any problem involving geometry, it is a good idea to *fill in the diagram with any relevant lines you can think of.* In this problem, it is natural to want to see the diameters of the circles.

Solution 10

When you draw the diameters through the centers, you see that the rectangle has dimensions d and $\frac{1}{2}d + d + \frac{1}{2}d = 2d$. Therefore, the area of the rectangle is $2d^2$. We have to subtract the areas of the circles within the rectangle. You should realize that these circular areas consist of one full circle and two semicircles, making the equivalent of two full circles. The area of a circle is πr^2, but r, the radius, is $d/2$. In terms of d, the area of two full circles is

$$2\pi \left(\frac{d}{2}\right)^2 \quad \text{or} \quad \frac{2\pi d^2}{4} \quad \text{or} \quad \frac{\pi d^2}{2}$$

The subtraction is

$$2d^2 - \frac{\pi d^2}{2} = d^2\left(2 - \frac{\pi}{2}\right)$$

or choice C.

Alternative Solution 10

By substituting an *even* number for the diameter, say, $d = 10$, you have that the rectangle has dimensions 10 and 20 and an area of 200. The radius of each of the circles is 5, and each circle has an area of 25π. The area of the shaded area is $200 - 50\pi$. Using $d = 10$ in each of the choices gives us

A $60 - 12\pi$ **B** $200 - 2\pi$ **C** $200 - 50\pi$ **D** $200 - 200\pi$
E $400 - 300\pi$

Choice C is our match.

Problems Involving Data and Statistics

Many problems involve reading and interpreting data from a table. In most cases this is a simple task. However, when some of the information in the table is in variable form, confusion may arise. Consider the following example.

Example 11

School Staff

Position	No. of staff	Annual salary, in thousands of $
Aide	6	c
Teacher	35	b
Supervisor	4	a

The table above represents the distribution of positions and annual salaries in a school. If $a = b + 12$ and $b = 3c$, what is the average salary, in thousands of dollars, of these school employees in terms of a?

A $\dfrac{42a - 456}{45}$ **B** $14a - 152$ **C** $\dfrac{a + b + c}{3}$ **D** $15a$ **E** $\dfrac{4a + 12}{3}$

Solution 11

A table such as the one given that specifies how many of each different type are in the entire group is referred to as a *frequency table*. It answers the question, "How frequent is it that, within the list of all members of the set, you would find a member in this category?" In this problem, it also tells you how frequently you can find the different salaries a, b, and c.

When finding an average, you have to account for all the members of the set. Therefore, you must remember to add the second column to determine that there are 45 employees in the school. The total salary is $6c + 35b + 4a$. We must use the given information

$$a = b + 12 \qquad \text{and} \qquad b = 2c$$

to express b and c in terms of a. That is,

$$b = a - 12 \qquad \text{and} \qquad c = \frac{1}{2}b = \frac{1}{2}(a - 12)$$

32

The total salary, in terms of a, is

$$6\left(\frac{1}{2}\right)(a-12) + 35(a-12) + 4a = 3a - 36 + 35a - 420 + 4a$$

$$= 42a - 456$$

Therefore, the average salary is

$$\frac{42a - 456}{45}$$

or choice A.

Note that choice C is not a possible answer since the problem specifically states that the answer must be in terms of a. Choice C involves the variables b and c.

Alternative Solution 11

This is a perfect problem for substitution as there is much algebra involved. You must, however, be sure to satisfy the relationships between a, b, and c. Note that the variables are meant to represent thousands of dollars. That is, an aide makes c *thousand dollars*. Starting with a simple choice for c, say, $c = 3$, we have $b = 2 \times 3 = 6$ and $a = 6 + 12 = 18$. The total salary becomes $6 \times 3 + 35 \times 6 + 4 \times 18 = 300$. The average salary is

$$\frac{300}{45}$$

(Don't reduce this fraction as it may be unnecessary. Do so only if you can't find this fraction among the choices.) With $a = 18$, the choices become

A $\dfrac{300}{45}$ **B** 100   **C** $\dfrac{36}{3}$ **D** 270   **E** $\dfrac{84}{3}$

and we clearly see that choice A is our answer.

Problems Involving Probability

The questions concerning probability on standardized tests usually are within what is referred to as *simple probability*. The key fact is, *The probability of an event occurring is the* ratio of the number of ways in which the event can occur compared to the number of all possible outcomes. There are many advanced concepts of probability that you may have learned in math classes, but they are usually not studied by the general population and, therefore, won't appear on popular standardized tests.

The next problem will review how to calculate a simple probability and also show you what could happen when you substitute and find more than one of the choices matching your numerical answer.

Example 12

There are b black marbles, p purple marbles, and 6 yellow marbles unseen in a container. If a marble is drawn, what is the probability that it is either purple or black?

A 1   **B** $\dfrac{2}{3}$ **C** $\dfrac{2}{b+p+6}$ **D** $\dfrac{b+p}{3}$ **E** $\dfrac{b+p}{b+p+6}$

Solution 12

The total number of possible outcomes is the total number of marbles in the container, $b + p + 6$. The number of successful outcomes is $b + p$. Therefore, the probability is

$$\frac{b+p}{b+p+6}$$

or choice E.

This is not a difficult problem, if you understand probability and are comfortable working with variable expressions.

Alternative Solution 12

Suppose we want to substitute $b = 1$ and $p = 1$. There would then be 8 marbles in the container, and 2 would satisfy the conditions of the problem. The probability is $2/8$. The choices would be

A 1   **B** $\dfrac{2}{3}$ **C** $\dfrac{2}{8}$ **D** $\dfrac{2}{3}$ **E** $\dfrac{2}{8}$

34

We now have a dilemma: Choices A, B, and D can be eliminated, but how do we choose between choices C and E? The answer: Substitute again with different numbers.

Now let us use $b = 2$ and $p = 3$. The new probability is $5/11$. Choice C becomes $2/11$, and choice E becomes $5/11$, which is now seen to be the correct answer.

The following problems will give you additional practice with the strategy of substituting. Try to solve the problems first using both a traditional solution and then by substitution. Additional insights, tips, and techniques into various situations will be given in solutions. Be sure to study them even if you get the problems correct.

Additional Problems

1. The product of x and the sum of x and y is equivalent to which of the following?

   A $(x + x)y$ **B** $x^2 + xy$ **C** $2xy$ **D** $x + xy$ **E** $x^2 + y$

2. If

$$\frac{x + y}{x}$$

   is a fraction less than 1, which of the following statements must be true?

   (I) x and y are both less than 1.
   (II) x and y are both negative.
   (III) xy is negative.

   A I only   **B** II only   **C** III only   **D** I and III only   **E** I, II, and III

3. If w is 30% of x and x is 10% of y and y is 70% of z, then the ratio of w to z is which of the following?

   A $\dfrac{7}{3}$ **B** $\dfrac{7}{4}$ **C** 2.1   **D** 0.21   **E** 0.021

4. Jen and Joey started bicycling toward each other at the same time along the same straight path that was m miles long. If Jen was traveling at r miles per hour and Joey was traveling at s miles per hour, which of the following gives the number of minutes it took for them to meet?

   A $\dfrac{m}{r} + \dfrac{m}{s}$ **B** $60m \left(\dfrac{1}{r} + \dfrac{1}{s} \right)$ **C** $\dfrac{m}{r + s}$ **D** $\dfrac{60m}{r + s}$ **E** $60m(r + s)$

5. In determining the pricing structure of a monthly plan, a telephone service uses the following. For d dollars per month, the customer

35

receives x minutes as part of the plan and is charged y cents for each additional minute during off-peak hours and 10 cents more for each additional minute during peak hours. Assuming that $x < 500$, if the customer uses 500 minutes all during off-peak hours, which of the following represents the customer's monthly cost in dollars?

A $\dfrac{d + [(500 - x)y]}{100}$ **B** $d + (500 - x)y$ **C** $d + 500y$

D $d + xy$ **E** $d + y$

6. If Bob can paint a room in h hours and Steve can paint the same room in 30 minutes less than Bob, which of the following represents the number of hours that it would take both men to paint the room working together?

A $2h - \dfrac{1}{2}$ **B** $\dfrac{1}{2h - \dfrac{1}{2}}$ **C** $\dfrac{h\left(h - \dfrac{1}{2}\right)}{2h - \dfrac{1}{2}}$ **D** $2h - 30$ **E** $\dfrac{1}{2h - 30}$

7. A grandmother who is y years old is x years older than her granddaughter. Fifteen years ago, the granddaughter was $1/4$ the age of her grandmother. Which of the following is true?

A $y - x = 15$ **B** $\dfrac{y}{x} - 15 = \dfrac{1}{4}$ **C** $\dfrac{y - x}{y} - 15 = \dfrac{1}{4}$

D $3y - 4x = 45$ **E** $2y - x = 15$

8. Stacey has q quarters, d dimes, and p pennies. Jeffrey has 7 fewer quarters and 5 more dimes than Stacy. While Jeffrey has no pennies, he has 6 more nickels than Stacey has pennies. Which of the following represents the amount of money that Stacey has more than Jeffrey?

A $25q + 10d + p$ cents   **B** $95 - 4p$ cents
C $5d - 7q + p - 6$ cents   **D** $50d + 175q + 5p + 30$ cents
E $18q - 5d - 4p$ cents

9. On a line segment with endpoints A and B, C is the midpoint of AB, D is a point between A and C, and E is a point between C and B. If $AD = BE$, which of the following must be true?

(I) $DC = CE$
(II) $AE = BD$
(III) $AE = \dfrac{3}{4}AB$

A I only   **B** I and II only   **C** I and III only
D II and III only   **E** I, II, and III

10. Two lines, ABC and $DEFG$, are parallel. If $m\angle ABE = a°$, $m\angle CBF = b°$,

36

$m\angle BFE = c°$, and $m\angle BEF = 30°$, which of the following is equivalent to $a - b$?

A 30   **B** 60   **C** $150 - c$ **D** $30 + c$ **E** $30 - c$

11. In a container, there are r red, g green, and y yellow lollipops. There are 3 times as many yellow lollipops as there are red ones and 5 more green lollipops than yellow ones. If a lollipop is taken from the container at random, which of the following represents the probability that the lollipop is red or green?

A $\dfrac{4}{7}$ **B** $\dfrac{2}{3}$ **C** $\dfrac{4y + 5}{7y + 5}$ **D** $\dfrac{4r + 5}{7r + 5}$ **E** $\dfrac{2r + 5}{7r + 5}$

12. In a certain class of n students, the average score on the last test was s. If m of those students had an average of t, which of the following represents the average of the remaining students?

A $\dfrac{ns - mt}{n - m}$ **B** $\dfrac{s - t}{n - m}$ **C** $s - t$ **D** $ns - mt$ **E** $\dfrac{s}{n} - \dfrac{t}{m}$

Solutions to Additional Problems

Solution to Problem 1

When interpreting a verbal description of arithmetic operations, you have to find the words that link the operations. That is, you should identify the phrases that refer to the operations mentioned. In most cases this is of the form *"The operation of ... and ..."* In this problem you have *"The product of ... and ...,"* which would lead you to x times the sum of x and y. The same structure appears in the last part of this expression, and you have $x(x + y)$. This is not one of the choices, so you must now perform the operations to find an equivalent form. The distributive law is applied to get $x^2 + xy$, choice B.

Alternative Solution to Problem 1

Substituting $x = 2$ and $y = 3$ leads to the statement "The product of 2 and the sum of 2 and 3," which alleviates confusion that might arise by working with unknown variables. The numerical computation is $2(2 + 3) = 2 \times 5 = 10$. Using $x = 2$ and $y = 3$ in the choices, we have the following:

A 12   **B** 10   **C** 12   **D** 8   **E** 7

We clearly see that choice B is the correct answer.

This problem also illustrates how using 1 as a value when substituting can lead to a problem for you. For example, if you let $x = 1$ and $y = 2$, the problem is $1(1 + 2) = 3$, and the choices are the following:

A 4   **B** 3   **C** 4   **D** 3   **E** 3

You could eliminate choices A and C, but then you would have to substitute again to determine which of the remaining three choices is correct.

Solution to Problem 2

The expression is equivalent to

$$\frac{x}{x} + \frac{y}{x} = 1 + \frac{y}{x}$$

For this to be less than 1, y/x must be a negative number. Clearly, statement III must be true. (Note the key word *must*!) This also implies that II cannot be true since a negative number divided by a negative number has a positive quotient. As for statement I, if x and y were two positive fractions both less than 1, y/x would remain a positive number and would not satisfy the conditions of the problem. Therefore, the only statement that must be true is III and choice C is our answer.

Alternative Solution to Problem 2

Before substituting, you have to read the three statements to get a sense of appropriate numbers to substitute. *In roman numeral problems, it is often the case that you have to substitute more than once to eliminate statements.* Since the statements mention negative numbers, let us substitute $x = -2$ and $y = -3$. Then

$$\frac{x + y}{x}$$

becomes

$$\frac{(^-2 + {}^-3)}{-2} = +\frac{5}{2}$$

which is greater than 1 and contradicts the given information. Therefore, statement II is not the case. We also see that statement I is not the case since both -2 and -3 are less than 1. Therefore, we can eliminate the four choices A, B, D, and E. We need only check statement III to make sure that we haven't done something wrong and eliminated more than we should have.

We can use $x = -2$ and $y = +3$. The given expression becomes

$$\frac{(^-2 + {}^+3)}{-2} = -\frac{1}{2}$$

which does satisfy the given information, and the product xy is -6, which makes statement III true. Therefore, choice C is our answer.

Solution to Problem 3

The key to the problem is to arrive at an equation that relates w and z. Using the decimal equivalents of the percentages, we have the following:

$$w = 0.3x \qquad x = 0.1y \qquad y = 0.7z$$

When replacing x with $0.1y$ and y with $0.7z$, we arrive at

$$w = 0.3(0.1y) = 0.03y = 0.03(0.7z) = 0.021z$$

Therefore, the ratio w/z is 0.021, choice E.

Alternative Solution to Problem 3

Since w is given in terms of x, which is given in terms of y, which is given in terms of z, we should substitute for z. The best value when working with percentages is $z = 100$. Therefore, y is 70% of 100 or 70, x is 10% of 70 or 7, and w is 30% of $7 = 0.3 \times 7 = 2.1$. Therefore, the ratio w/z is

$$\frac{2.1}{100} = 0.021$$

or choice E.

Solution to Problem 4

The key to the problem is using the formula distance = rate × time, and the fact that when Jen and Joey meet, they had covered the total distance of m miles and had traveled the same amount of time. Let t represent the time traveled. Jen's distance is rt, and Joey's distance is st. We now can create the equation $rt + st = m$, or $t(r + s) = m$. Dividing both sides by $r + s$ gives us

$$t = \frac{m}{r + s}$$

This solution, however, is in hours since the rates were given in miles per hour. The number of minutes in t hours is $60t$. Therefore, the number of minutes is

$$\frac{60m}{r + s}$$

or choice D.

Alternative Solution to Problem 4

The facts remain the same as mentioned above. Let's substitute with relatively easy numbers such as $m = 10$, $r = 3$, and $s = 7$. The problem then becomes simple. That is, with Jen traveling at 3 miles per hour and Joey traveling at 7 miles per hour, they would need only 1 hour to cover the entire distance of 10 miles. Therefore, we should be looking for the choice that gives 60 minutes as the answer. With $m = 10$, $r = 3$, and $s = 7$, the choices become the following:

A $\dfrac{10}{3} + \dfrac{10}{7}$ **B** $\dfrac{600}{\left(\dfrac{1}{3} + \dfrac{1}{7}\right)}$ **C** $\dfrac{10}{10}$ **D** $\dfrac{600}{10}$ **E** $600(10)$

Clearly, choice D is the only one that produces 60 as an answer.

Solution to Problem 5

There is much going on in this problem, and you have to carefully determine which information is relevant. Notice that all the calls are assumed to have been made during off-peak hours, which renders as irrelevant the information about an additional 10 cents. Therefore, we are working with the given x minutes and y cents for each of the remaining $500 - x$ minutes. From that we know the total monthly charge is d dollars $+ (500 - x)y$ cents. The question explicitly states that the answer be in dollars. Cents are converted to dollars by dividing by 100. Therefore, the charge in dollars is

$$\frac{d + [(500 - x)y]}{100}$$

or choice A.

Alternative Solution to Problem 5

Substituting simple values such as $d = 20$ dollars, $x = 100$ minutes, and $y = 5$ cents, the problem becomes the following:

> For 20 dollars per month, the customer receives 100 minutes as part of the plan and is charged 5 cents for each additional minute during off-peak hours and 15 cents for each additional minute during peak hours. If the customer uses 500 minutes all during off-peak hours, what is the monthly cost in dollars?

This statement is much simpler to understand, and it is easy to make the computation. The monthly charge would be $\$20 + 400 \times \$0.05 = \$20 + \$20 = \$40$. With $d = 20$, $x = 100$, and $y = 5$, the choices become the following:

40

A $20 + \dfrac{400 \times 5}{100} = 20 + 20 = 40$ **B** $20 + 400 \times 5 = 2{,}020$

C $20 + 500 \times 5 = 2{,}520$ **D** $20 + 100 \times 5 = 520$ **E** $20 + 5 = 25$

Clearly, the correct choice is A.

Solution to Problem 6

(The direct solution to this problem will test your algebra skills to the fullest!) The key formula to apply in a problem relating to work being done is the following:

Number of jobs completed = rate of work × time to complete the task

The nature of this problem is similar to problem 4 above in that you can imagine both men starting at the same point and painting the walls around the room in opposite directions until they meet. They will both have used the same time. We need to express their rates of work, and we can derive the answer from the following:

I completed job = Bob's rate × time + Steve's rate × time

Bob's rate is given as 1 job per h hours or $1/h$ jobs per hour. Steve's rate is given as 1 job per $(h - \frac{1}{2})$ hours or

$$\dfrac{1}{\left(h - \dfrac{1}{2}\right)} \text{ jobs per hour}$$

(Note that we have to convert 30 minutes into $\frac{1}{2}$ hour.) Therefore, letting t be the number of hours, we have the following equation:

$$1 = \frac{1}{h} \times t + \frac{1}{\left(h - \dfrac{1}{2}\right)} \times t \quad \text{or} \quad 1 = \left[\frac{1}{h} + \frac{1}{\left(h - \dfrac{1}{2}\right)}\right] t$$

The expression in parentheses needs to be combined into a single fraction with a common denominator. This would be $h(h - \frac{1}{2})$, and it becomes

$$\dfrac{\left(h - \dfrac{1}{2} + h\right)}{h\left(h - \dfrac{1}{2}\right)} \quad \text{or} \quad \dfrac{\left(2h - \dfrac{1}{2}\right)}{h\left(h - \dfrac{1}{2}\right)}$$

To solve the equation, we would divide both sides of the equation by this term to get the following:

$$t = \cfrac{1}{\cfrac{2h - \dfrac{1}{2}}{h\left(h - \dfrac{1}{2}\right)}} = \cfrac{h\left(h - \dfrac{1}{2}\right)}{2h - \dfrac{1}{2}}$$

or choice C.

Alternative Solution to Problem 6

Even with substitution, you still have to apply the same formula and recognize that the number of jobs completed is 1, which is the sum of both of their efforts over the same amount of time. Let us suppose Bob can paint the room in 2 hours for a rate of $1/2$ jobs per hour. Therefore, Steve can paint the room in 1.5 hours for a rate of $1/1.5$ or $2/3$ jobs per hour. We still need a little algebra here with t being the time. The equation is as follows:

$$1 = \frac{1}{2}t + \frac{2}{3}t = \left(\frac{1}{2} + \frac{2}{3}\right)t = \frac{7}{6}t$$

Therefore,

$$t = \frac{1}{\dfrac{7}{6}} \quad \text{or} \quad t = \frac{6}{7}$$

(Note that $1/*$ is equal to the *reciprocal* of $*$). If you are using a calculator, you may want to use the decimal equivalent of $6/7$, which is approximately 0.857.

Using $h = 2$, the choices become the following:

A 3.5   **B** $\dfrac{1}{3.5} = \dfrac{2}{7} = 0.286$ **C** $\dfrac{2 \times 1.5}{3.5} = 0.857$

D $4 - 30 = -26$ **E** $\dfrac{1}{-26}$

You can easily recognize choice C as the correct answer.

Solution to Problem 7

The current age of the grandmother is y years old, and the current age of the granddaughter is $y - x$ years old. Fifteen years ago, the grandmother was $y - 15$ years old and the granddaughter was $y - x - 15$ years old. The

42

ratio of the granddaughter's age to the grandmother's age 15 years ago was

$$\frac{y - x - 15}{y - 15} = \frac{1}{4}$$

Cross multiplying, we have $4y - 4x - 60 = y - 15$. Transposing y to the left and -60 to the right, we have the expression $3y - 4x = 45$, or choice D.

Alternative Solution to Problem 7

In order to substitute effectively, you ought to select ages that satisfy the ratio for 15 years ago. Suppose the grandmother was 80 years old and the granddaughter was 20 years old 15 years ago, which gives the ratio $1/4$. This implies that their current ages are 95 years old and 35 years old. Therefore, $y = 95$ and $x = 60$, the difference between their ages. With these values, you can examine the validity of each equation:

A $95 - 60 = 15$ (false)   **B** $\frac{95}{60} - 15 = \frac{1}{4}$ (false)

C $\frac{35}{95} - 15 = \frac{1}{4}$ (false)   **D** $3(95) - 4(60) = 285 - 240 = 45$ (true)

E $2(95) - 60 = 15$ (false)

Clearly, choice D is the correct answer.

Solution to Problem 8

Using q, d, and p as given to represent the number of each coin that Stacey has, we determine that Jeffrey has $q - 7$ quarters, $d + 5$ dimes, and $p + 6$ nickels. Therefore, using cents to calculate each person's amount, we have that Stacey has $25q + 10d + p$ cents and Jeffrey has

$$25(q - 7) + 10(d + 5) + 5(p + 6) = 25q - 175 + 10d + 50 + 5p + 30$$
$$= 25q + 10d + 5p - 95 \text{ cents}$$

Subtracting Jeffrey's amount from Stacey's, we have the expression $95 - 4p$ cents, choice B.

Alternative Solution to Problem 8

Suppose we substitute with $q = 8$, $d = 5$, $p = 4$, and $n = 3$. This means that Stacey has 8 quarters, 5 dimes, and 4 pennies, which means she has 254 cents. Jeffrey has 1 quarter, 10 dimes, and 10 nickels, which amounts to 175 cents. Stacey has 79 cents more than Jeffrey. With these values, the choices become the following:

A 254 cents **B** 79 cents **C** -127 cents **D** 4,675 cents **E** 409 cents

The correct answer is clearly choice B. *(This is much simpler than the algebra in the first solution!)*

Solution to Problem 9

The first step, without question, is to draw a diagram depicting the given situation as close to scale as you can. For this problem the diagram is

$$A \text{_____} D \text{_____} C \text{_____} E \text{_____} B$$

The unknowns in this problem are the length of all of AB—let's call this x—and the lengths of the equal segments AD and BE—let's call these lengths y.

Since C is the midpoint of AB, $AC = BC = \frac{1}{2} x$. You can find the lengths of DC and CE by subtraction. That is, $DC = AC - AD = \frac{1}{2} x - y$ and $CE = CB - BE = \frac{1}{2} x - y$. Therefore, we see that statement I is true, $DC = CE$.

To check statement II, we can subtract as follows:

$$AE = AB - BE = x - y \qquad \text{and} \qquad BD = AB - AD = x - y$$

Therefore, $AE = BD$, and statement II is true.

Statement III would be true only if $\frac{3}{4} x = x - y$, which would mean that $y = \frac{1}{4} x$. There is no condition in the problem that insists that this be the case. Therefore, statement III does not necessarily have to be true. Our answer is choice B.

Alternative Solution to Problem 9

It seems obvious that working with the variables x and y can add some confusion to the problem. We can, instead, substitute for the length of the line segment AB, say, 10, and for the equal smaller segments AD and BE, say, 2. Marking the diagram with these numbers and filling in with the obvious lengths for DC and EC, we have the following:

$$A \underset{2}{\text{\_\_\_}} D \underset{3}{\text{_____}} C \underset{3}{\text{_____}} E \underset{2}{\text{\_\_\_}} B$$

With these values, we see that statements I and II are true and statement III is not (8 is not $\frac{3}{4}$ of 10). You should try other values to feel more confident about your conclusion. *(Note that if you make all the line segments equal, statement III might appear to be true.)* Substituting quickly leads you to the correct answer, choice B.

44

Solution to Problem 10

The diagram for the picture must be drawn. Remember that when more than one point is used to describe a line, the points appear on the line in the order given. Therefore, the diagram is the following:

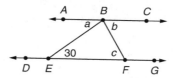

You should notice that there are two pairs of *alternate interior angles* within the diagram, and you should therefore use the key fact *If two parallel lines are cut by a transversal, then the angles formed interior to the parallel lines and on alternate sides of the transversal are congruent.* ∠*ABE* and ∠*BEF* are a pair of such angles, and, therefore, $a = 30$. ∠*BFE* and ∠*CBF* are the other pair of such angles, and, therefore, $b = c$. With this, we have $a - b = 30 - c$, or choice E.

Alternative Solution to Problem 10

In this problem, you are limited in the ways you can make substitutions as a result of the key fact mentioned above. Your only real variable is the value of *b*. (However, if you try to substitute for *a* and *c* as well, you might trigger your memory cells to enable you to recall information about alternate interior angles as the diagram would contradict what you know about parallel lines.) Recognizing that $a = 30$ and $b = c$, if you let $b = 50$, you will have $c = 50$ and $a - b = 20$. The choices become the following:

A 30   **B** 60   **C** 100   **D** 80   **E** 20

Clearly, choice E is the correct answer.

Solution to Problem 11

The probability of selecting a red or green lollipop will be the number of red and green lollipops in the container divided by the total number of lollipops in the container. That is,

$$\frac{r + g}{r + g + y}$$

The given information tells us that $y = 3r$ and $g = y + 5 = 3r + 5$. Therefore, the probability is the following:

$$\frac{r + 3r + 5}{r + 3r + 5 + 3r} = \frac{4r + 5}{7r + 5}$$

or choice D.

Alternative Solution to Problem 11

The given information indicates that if we have a value for r, we can find the value for y and, then, the value for g. Let's use a value for r that avoids other numbers mentioned in the problem. Suppose $r = 8$. Then $y = 24$ and $g = 29$. The probability is, therefore, $37/61$. With these values for r, y, and g, the choices become the following:

A $\dfrac{4}{7}$ **B** $\dfrac{2}{3}$ **C** $\dfrac{101}{173}$ **D** $\dfrac{37}{61}$ **E** $\dfrac{21}{61}$

Clearly, choice D is the correct answer.

Solution to Problem 12

If the average of n students is s, then the sum of all the students' scores is ns. The sum of the scores of the m students whose average is t is mt. Therefore, the sum of the scores of the remaining $n - m$ students is $ns - mt$. Their average is as follows:

$$\frac{ns - mt}{n - m}$$

or choice A.

Alternative Solution to Problem 12

Substituting for n, m, s, and t will make the problem simpler. In fact, you can simplify the situation further by assuming that the average score for the m students was the actual score for all of them and the same is true for the remaining students, if you let m be one-half the class. That is, let $n = 8$ be the number of students in the class, and let $s = 80$ be the average. Let $m = 4$ and $t = 100$ be their average. Assume that all four of these students each scored 100 on the test. The remaining four students could have each scored a 60 so that the class average would indeed be 80. Thus, the average of the remaining four would be 60. With $n = 8, s = 100$, $m = 4$, and $t = 100$, the choices become the following:

A $\dfrac{640 - 400}{8 - 4} = \dfrac{240}{4} = 60$ **B** $\dfrac{80 - 100}{8 - 4} = \dfrac{-20}{4} = -5$

C $80 - 100 = -20$ **D** $640 - 400 = 240$

E $\dfrac{80}{8} - \dfrac{100}{4} = 10 - 25 = -15$

Choice A is clearly the correct answer.

Chapter 3

Using Ratios and Proportions Quickly

Some word problems set up situations in which you have to compare rates. Such problems might ask about who completes a job faster, which vehicle reaches a location first, which activity will cost more money, and so on. In all these situations, you will need to examine the base rates of the people, machines, or activities involved. If you are familiar with the algebraic solutions to problems involving ratios and proportions, these problems may be easy for you. With word problems, however, the base rates may be hidden in the problem. The strategies in this chapter will help you uncover them quickly.

Ratios

A *ratio* is a comparison of two quantities. For example, the statement "In a class the ratio of boys to girls is 3:2" means that if you divided the number of boys by the number of girls, you would have $^3/_2$, or 1.5. You could, therefore, also say that the ratio of boys to girls is 1.5:1. However, we usually do not use decimals within a ratio.

The ratio does not tell us the total number of objects in the whole collection. In the above example, if the class had exactly 5 students, there would be exactly 3 boys and 2 girls. However, if the class had 10 students, there would be exactly 6 boys and 4 girls to maintain the ratio of 3:2. In this case, we see that there is a multiplier, 2, needed to determine the exact numbers of boys and girls. If the class had 20 students, the multiplier would be 4, and there would be 12 girls and 8 boys. If the class had 50 students, the multiplier would be 10, and there would be 30 girls and 20 boys. When the total number of objects is not given directly but is needed in the problem, the algebraic approach to the problem is to let x be the multiplier. In our example, the number of boys would be $3x$, the number of girls would be $2x$, and the total number of students in the class would be $3x + 2x = 5x$. Let us use this idea in the following examples.

Example I

In a container holding only blue and red marbles, the ratio of red marbles to blue marbles is 7:4. Which of the following could be the total number of marbles in the container?

A 18 **B** 19 **C** 20 **D** 21 **E** 22

Solution I

If there were only 11 marbles, there would be exactly 7 red marbles and 4 blue marbles. However, 11 is not one of the choices. We need to use a multiplier, which must be a whole number since we cannot have fractions of a marble. This implies that the total number must be a multiple of 11. The only one of the choices that satisfies this condition is choice E, 22. (There would be exactly 14 red marbles and 8 blue marbles.)

Alternative Solution 1

You could make use of what you learned in Chapter 1 and work backward with each choice. Let x be the multiplier. This implies that there are $7x$ red marbles, $4x$ blue marbles, and $11x$ marbles all together. Starting with choice C, you have the equation $11x = 20$ and $x = 20/11$. Clearly, this cannot be correct since x must be a whole number. Choice E becomes the only acceptable choice.

There is another way to use the choices for this problem. Suppose you started with choice C and assumed that there were 20 marbles in the container. The problem would be to find two numbers whose sum is 20 and that are in the ratio of 7:4. This would mean listing all the possible ways of doing so: 19 and 1, 18 and 2, 17 and 3, and so on. When you reach 14 and 6, you might gain some insight into the problem in the following way. If there were 22 marbles, you would have 14 and 8, which gives the ratio 7:4. If your choices were larger numbers, this strategy would be too time-consuming to be worthwhile.

Example 2

A ranch has cattle and horses in a ratio of 9:5. If there are 80 more head of cattle than horses, how many animals are on the ranch?

A 140   **B** 168   **C** 238   **D** 280   **E** 308

Solution 2

Algebraically, let x be the multiplier. Therefore, there are $9x$ head of cattle, $5x$ horses, and $14x$ animals all together. The second condition of the problem tells us that $9x = 5x + 80$. This gives us $4x = 80$ and $x = 20$. There are $9 \times 20 = 180$ head of cattle, $5 \times 20 = 100$ horses, and 280 animals all together, choice D.

Alternative Solution 2

You should realize that the total number has to be a multiple of 14. By using the choices, we can divide by 14 to find the multiplier. You would determine the number of head of cattle and the number of horses and check if the difference of these numbers was 80. You might be inclined to start with choice A as it quickly implies that the multiplier would be 10. That would imply that there were 90 head of cattle and 50 horses. However, $90 - 50 = 40$. You should realize that you need twice as many animals to satisfy the problem, 280, choice D.

Example 3

A paint mixture contains b parts of blue paint and y parts of yellow paint. If g gallons of this mixture are needed, which of the following represents the amount of blue in the mixture?

A $\dfrac{b}{g}$ **B** $\dfrac{bg}{b+y}$ **C** $\dfrac{b}{g(b+y)}$ **D** $\dfrac{g-y}{g}$ **E** $\dfrac{g-y}{b+y}$

Solution 3

Parts is another way of indicating a ratio without specifying a total amount. In this problem, we have that the ratio of the amount of blue to the amount of yellow is b/y. If we let the multiplier be x, we should realize that x is a measure of gallons and that it need not be a whole number as we can have fractions of a gallon. There are bx gallons of blue in the mixture and yx gallons of yellow in the mixture. The total number of gallons is $bx + yx = g$. We can solve for x as follows:

$$x(b + y) = g \qquad \text{and} \qquad x = \frac{g}{b + y}$$

Therefore, there are

$$bx = \frac{bg}{b + y} \text{ gallons of blue in the mixture}$$

or choice B.

Alternative Solution 3

Using substitution, as you learned in Chapter 2, will simplify this problem. Suppose $b = 3$, $y = 2$, and $g = 5$. There are then exactly 3 gallons of blue and 2 gallons of yellow in the 5-gallon mixture. With these values, the choices become the following:

A $\dfrac{3}{5}$ **B** $\dfrac{15}{5} = 3$ **C** $\dfrac{3}{25}$ **D** $\dfrac{3}{5}$ **E** $\dfrac{3}{5}$

Clearly, the correct answer is B.

Rates and Proportions

Many word problems involve situations in which *similar events* or *similar shapes* occur. The word *similar* in everyday language often means that there are common aspects of two situations, but they have enough different aspects to keep them unequal. In mathematics, *similar* has a specific meaning related to ratios. Let us use the following definitions:

Two events are similar if they are dependent on equal rates of activity.

Two shapes are similar if each angle of one has a corresponding congruent angle in the other and all the ratios of corresponding sides are equal.

Another word needs to be defined in order to help us discuss similarity in this way, *proportion:*

A proportion is an equation that states that two ratios are equal.

That is,

$$a:b = c:d \quad \text{or} \quad \frac{a}{b} = \frac{c}{d}$$

The outer values, a and d, are called the *extremes of the proportion*, and the inner values, b and c, are called the *means of the proportion.* Notice that these equal ratios give rise to another equation involving the *cross products* around the equal sign.

$$bc = ad$$

That is, *the product of the means is equal to the product of the extremes.*

On standardized tests, the key to correctly solving word problems involving similar events is to quickly identify the equal rates or ratios and to set up the correct proportion. In most cases one of the four quantities will be an unknown, which can be solved by using the equation involving the cross products. The following problems will demonstrate this and will also present an alternative strategy.

Example 4

Alison and Christine are each taking multiple-choice tests and work at the same pace. Alison's test has 20 multiple-choice questions while Christine's test has 25 questions. How many minutes did Alison take to complete her test if Christine took 40 minutes to complete her test?

A 20   **B** 28   **C** 30   **D** 32   **E** 40

Solution 4

The key phrase in the problem is *at the same pace*, which means that they worked at the same *rate*. The rate of work is the ratio number of minutes divided by number of questions, which we would express as minutes per question. The proportion would come from the fact that Alison's rate = Christine's rate. If we let m be the number of minutes that Alison took to complete the test, we have the proportion

$$\frac{20}{m} = \frac{25}{40}$$

The cross-product equation is $25m = 800$. Dividing both sides by 25 gives us $m = 32$, choice D.

Alternative Solution 4

If we think of the problem in the following way, we can develop an alternative solution:

If it takes 40 minutes to answer 25 questions, how long will it take to answer 20 questions?

This is the essence of the problem since the girls are working at the same pace.

We can answer this question by finding out how long it takes to answer 1 question at the given rate and then multiply that *unit rate* by 20. Using arrows to show the cause-and-effect relationship between the number of questions and the number of minutes is helpful to understand this:

questions → minutes
25 → 40

If the rate is always the same, we can divide on both sides of the arrow by 25 to find the unit rate.

$$1 \rightarrow \frac{40}{25} = 1.6$$
$$20 \rightarrow 1.6 \times 20 = 32$$

You can see that once you know the unit rate, you can determine the number of minutes for any number of questions. Let's call this strategy *using arrows*.

This alternative solution is an example of what is called *proportional reasoning,* and, in fact, it is the method many people use to solve problems like this in the real world when paper and pencil are not available. The following problem demonstrates where this strategy will help to simplify the problem.

Example 5

A recipe calls for 1 cup of milk for every $2\frac{1}{2}$ cups of flour to make a cake that would feed 6 people. How many cups of both flour and milk will need to be measured to make a similar cake for 8 people?

A $1\frac{1}{3}$ **B** $3\frac{1}{3}$ **C** $4\frac{2}{3}$ **D** $5\frac{1}{2}$ **E** 7

Solution 5

The word *similar* indicates that you should look for a proportion involving equal ratios. However, there are three different objects in this problem creating a three-part ratio, milk : flour : people. If we let *m* be the number of cups of milk for 8 people and *f* be the number of cups of flour for 8 people, we have $1:2\frac{1}{2}:6 = m:f:8$. However, in order to use a proportion, we can include a ratio of only two of these objects. We have three possibilities:

milk / flour   milk / people   flour / people

Since we don't know either of the new amounts for milk or flour, we need to use one of the ratios involving people to start.

If we focus first on milk, we have the following proportion:

$$\frac{m}{8} = \frac{1}{6}$$

The cross products give us the following:

$$6m = 8 \quad \text{and} \quad m = \frac{8}{6} \quad \text{or} \quad 1\frac{1}{3} \text{ cups}$$

For flour, we use

$$\frac{f}{8} = \frac{2\frac{1}{2}}{6}$$

leading to

$$6f = 20 \quad \text{and} \quad f = \frac{20}{6} = 3\frac{1}{3} \text{ cups}$$

Therefore, the number of cups needing to be measured for milk and flour is as follows:

$$1\frac{1}{3} + 3\frac{1}{3} = 4\frac{2}{3}$$

or choice C.

Alternative Solution 5

Using arrows, we set up the following scheme:

$$\text{people} \rightarrow \text{milk} \rightarrow \text{flour}$$
$$6 \rightarrow 1 \quad \rightarrow 2\frac{1}{2}$$
$$1 \rightarrow \frac{1}{6} \quad \rightarrow 2\frac{1}{2} \div 6 = \frac{5}{12}$$
$$8 \rightarrow \frac{8}{6} \quad \rightarrow \frac{40}{12} = \frac{20}{6}$$

Therefore, we need

$$\frac{8}{6} + \frac{20}{6} = \frac{28}{6} = 4\frac{2}{3}$$

cups of both milk and flour.

There is a way to avoid the cumbersome fractions in the third line of the scheme. Rather than going to a unit rate, think ahead of how you can get from 6 to 8 to clear the fraction for the number of cups of flour. Calculating the situation for

12 people will do this as follows:

$$\text{people} \rightarrow \text{milk} \rightarrow \text{flour}$$
$$6 \rightarrow 1 \quad \rightarrow 2\frac{1}{2}$$
$$12 \rightarrow 2 \quad \rightarrow 5$$
$$1 \rightarrow \frac{2}{12} \quad \rightarrow \frac{5}{12}$$
$$8 \rightarrow \frac{16}{12} \quad \rightarrow \frac{40}{12}$$

The sum we need is now as follows:

$$\frac{16}{12} + \frac{40}{12} = \frac{56}{12} = 4\frac{8}{12} = 4\frac{2}{3}$$

If you are uncomfortable with algebra, you should practice this strategy so that you can apply it quickly.

Example 6

Two hexagons, A and B, are similar, and the ratio of corresponding sides of A to B is 4:3. If the sides of hexagon A are each halved while the sides of hexagon B are doubled, what will be the new ratio of the corresponding perimeters of the larger hexagon to the smaller hexagon?

A 1:3 **B** 2:3 **C** 1:1 **D** 3:2 **E** 3:1

Solution 6

Once again, a multiplier, x, can be used to allow us to represent the actual quantity, in this case a side of the hexagon. Let a pair of corresponding sides of the hexagons be $4x$ and $3x$. Halving the side of hexagon A and doubling the side of hexagon B gives us the new sizes of these corresponding sides, $2x$ and $6x$. Note that the problem did not state that the hexagons are *regular*, which would mean that all the sides are equal in length. Since a hexagon has six sides, we would have to do the same for each of the six. However, the perimeters of similar polygons are in the same ratio as the corresponding sides. Therefore, the ratio of the larger perimeter to the smaller perimeter would be 6:2, or 3:1, choice E.

Alternative Solution 6

You can *substitute* lengths for the sides of each hexagon, remembering to keep the ratio. Therefore, you can use multiples of 4 for the sides of hexagon *A* and multiples of 3 for the sides of hexagon *B*. Suppose hexagon *A* had sides of 4, 8, 12, 16, 20, and 24, and the corresponding sides of hexagon *B* had sides of 3, 6, 9, 12, 15, and 18. The new sides of hexagon *A* would be 2, 4, 6, 8, 10, and 12, and its perimeter would be 42. The new sides of hexagon *B* would be 6, 12, 18, 24, 30, and 36, and its perimeter would be 126. The ratio of the larger perimeter to the smaller would be 126:42 or 3:1.

After reading the last paragraph, you can see that the perimeter follows the same pattern as each of the sides when the same kind of change is made to all of them. That is, when the sides are each halved, the perimeter must be halved as well. It should also be evident that *the perimeters of two similar polygons should be in the same ratio as the corresponding sides.*

When considering areas, however, something different occurs as the next problem indicates.

Example 7

Three triangles are similar, and their sides are in the ratio of 2:5:6. If each side of the smallest triangle is tripled while each side of the largest triangle is divided by 3, what is the ratio of the areas of the three triangles after these changes are made?

A 6:5:2 **B** 36:25:4 **C** 216:125:8

D 18:15:6 **E** 108:75:12

Solution 7

A straightforward algebraic solution is impossible to put together for this problem using the traditional formula *area* $= \frac{1}{2} \times base \times height$ since we do not know anything about the heights of the triangles. (There is a way to find the area of a triangle by using the lengths of the three sides, *Heron's Formula*, which is beyond the scope of this book.) However, we can use an algebraic technique to understand the relationship between the areas of similar triangles.

Suppose the base and the height of a triangle are *b* and *h*. To represent the corresponding base and height of a similar triangle, we use a multiplier x, and have bx and hx. That is, the ratio of the sides of the larger triangle to the smaller triangle is x, or, in other words, the second triangle is x times as large in length (or perimeter) as the first.

The area of the first triangle is $1/2 \times b \times h$, and the area of the second is $1/2 \times bx \times hx$ or $1/2 \times b \times h \times x^2$. Therefore, the ratio of the areas of the second triangle to the first is x^2. In other words, the second triangle is x^2 times as large as the first in area. The key fact to remember is *The ratio of the areas of two similar triangles is the square of the ratio of the lengths of the sides.*

With this fact, the answer to the problem is obvious. If the ratio of the lengths of the sides of the three triangles is 2:5:6, then after tripling each side of the smallest triangle and reducing the length of each side of the larger triangle by a factor of 3, the new ratio is 6:5:2. The ratio of the areas is, therefore, 36:25:4, choice B.

Alternative Solution 7

A useful strategy to solve a problem is to try a simpler situation. (Let's call it *make it simple.*) This sometimes involves making certain assumptions that may or may not be true about the situation, but the strategy is intended to let you handle a situation you are familiar with and see if it gives you some insight.

In this problem, a simpler situation would arise if you assumed that all three triangles were right triangles. In fact, you can use the familiar 3-4-5 right triangle as the structure of all three triangles. Therefore, the smallest triangle would have sides of 6, 8, and 10, the second triangle would have sides of 15, 20, and 25, and the third triangle would have sides of 18, 24, and 30. Tripling the sides of the first triangle would make its sides 18, 24, and 30, and reducing the sides of the third triangle by a factor of 3 would make its sides 6, 8, and 10

The area of a right triangle is simply $1/2 \times$ leg $\times$ leg, since one leg can be a base and the other, since it is perpendicular to the first, would be the height of the triangle. Therefore, the areas of the triangles are $1/2 \times 18 \times 24 = 216$, $1/2 \times 15 \times 20 = 150$, and $1/2 \times 6 \times 8 = 24$. The ratios of the areas of the triangles are 216:150:24, or, when reduced by the common factor of 6, 36:25:4.

We have seen in the above two examples that the *perimeters* of similar figures are in the same ratio as the *lengths* of the corresponding sides while the *areas* are in the same ratio as the *squares* of the lengths of the corresponding sides. There is also a related fact about the volumes of similar three-dimensional shapes. That is, the ratio of the *volumes* of similar three-dimensional shapes are in the same ratio as the *cubes* (or third powers) of the lengths of the corresponding sides.

Percentages

Percentages, as we have seen in the past, give us a way to talk about fractions, and they also play a role in determining cost, pricing, and wages. Another way to think about a percentage is as a ratio whose second compared number is 100. That is, 35% is 35:100. As we've seen in previous problems with ratios, this does not mean that the actual numbers are 35 and 100. Percentages give us a way to communicate about ratios in a uniform way. For example, two people could be talking about what they believe are different situations, and one states that the ratio he is studying is 6:25 while the other says that the ratio is 4:15. Who has the higher ratio? By converting each to a percentage, or a ratio with a common second number of 100 for comparison, the answer is easily obtained. That is,

$$\frac{6}{25} = 0.24 = 24\% \quad \text{and} \quad \frac{4}{15} = 0.2666\ldots$$
$$= 26.7\%$$

Clearly, the second ratio is larger.

Example 8

A library contains three times as many novels as all other books combined. What percentage of the library are all the other books?

A 10% **B** 25% **C** $33\frac{1}{3}\%$ **D** 75%

E There is insufficient information to determine the answer.

Solution 8

At first, you might wonder why you aren't given the total number of books in the library. This would be relevant information, and you might be quick to believe the answer is E. However, it is not necessary since the problem is talking about a ratio rather than an exact number. An algebraic solution is to let x be the number of all other books, which means that $3x$ is the number of novels. The total number is $4x$. The ratio we are looking for is $x/4x$, or $1/4$, which is 25%, choice B.

Alternative Solution 8

Since you are working with percentages, you can solve the problem by assuming that there are 100 books. With this in mind, you might even realize instantaneously that there would have to be 25 other books and 75 novels to meet the conditions of the problem. If the answer did not come quickly, you could use the choices to determine which would satisfy the problem. That is, with choice A, there would be 10 other books and 90 novels, which is clearly not the case for the situation. Moving to choice B immediately gives the answer.

Note that choice C would be difficult to use with 100 books since you can't have a fraction of a book. If this were the case, you would have to change the number you were considering. That is, since $33\frac{1}{3}$% is the fraction $1/3$ (a good point to remember), think of having 300 books. That would mean that there are $1/3$ of $300 = 100$ other books and, therefore, 200 novels. This clearly does not satisfy the conditions of the problem.

When faced with a problem whose choices contain a statement like choice E in the previous example, you have to remember to consider all possibilities that might occur. One of the other choices might be correct in one case, while another choice can be correct in a different case. If two different answers arise, the answer must be E. (In Chapter 5 we will study problems where the choices always contain this possibility.)

Example 9

At one point in the season, a basketball team's performance indicates that its win-loss ratio is 45%. If its win-loss ratio for the next 38 games is 90%, what is its overall win-loss ratio?

A 45%   **B** 49.1%   **C** 52.5%   **D** 135%

E There is insufficient information to determine the answer.

Solution 9

You must be careful to read the problem and understand what is being told to you. Most often, we are given *winning percentages*, which are the ratios of wins to total number of games played. This problem talks about win-loss ratio. Note that if there is an equal number of wins and losses, the ratio is 1:1 or 100%. Having more wins than losses would give a percentage greater than 100%.

Attempting to solve the problem algebraically requires assigning variables to represent the number of games won and lost to result in a win-loss ratio of 45%. Let x be the number of games won and y be the number of games lost. The problem tells you that $x/y = 45/100$.

Now you have to determine what has actually happened over the next 38 games. If the win-loss ratio is 90%, or 90/100, you can reduce that result to 9/10 and interpret it to mean that if 19 games have been played, the team would have won 9 and lost 10. Since 38 is twice 19, the team would have won 18 and lost 20.

Therefore, the total number of wins for the team was $x + 18$, and the total number of losses was $y + 20$. The new win-loss ratio would be

$$\frac{x + 18}{y + 20}$$

This is where you have to stop. Since you don't know the values of x and y, you cannot determine the actual number. Therefore, the correct choice is E.

Alternative Solution 9

By considering the choices, we can quickly determine that choices A and D can be eliminated. We must still determine that in the next 38 games, there were 18 wins and 20 losses. Choice A, 45%, can be easily eliminated since the win-loss ratio must rise as the team added more wins and fewer losses. Choice D, 135%, would be possible only if there were more total wins than losses. This can't happen because in both the original number of games and in the next 38, the win-loss percentages are under 100%, indicating fewer wins than losses.

Having a choice such as E requires us to consider all possibilities and to determine if more than one possible situation can exist. Suppose there were exactly 45 wins and 100 losses in the original set of games. With the next 38 games, we would have $45 + 18 = 63$ wins and $100 + 20 = 120$ losses. The win-loss ratio would be $63/120 = 52.5\%$. We might want to say that choice C is the answer and move on to the next question. However, if the original set had 90 wins and 200 losses, which also gives a 45% win-loss ratio, the new win-loss ratio would be

$$\frac{108}{220} = 49\frac{1}{11}\%$$

or approximately 49.1%, which is choice B! Therefore, the answer must be E.

As you work on the additional problems that follow, you will see more situations in which rates, ratios, and proportions arise.

Additional Problems

1. In a popular fruit drink the ratio of fruit juice to water is 3:5. How many ounces of fruit juice are present in a 2-quart pitcher of the drink?

   A 3   **B** 6   **C** 12   **D** 24   **E** 32

2. The prices of two coats are in the ratio of 7:5. If the sum of the

costs of both coats is d dollars, what is the positive difference of the costs of the coats?

A $\dfrac{d}{12}$ **B** $\dfrac{d}{6}$ **C** $\dfrac{2d}{7}$ **D** $\dfrac{2d}{5}$ **E** $\dfrac{5}{7}d$

3. The average person can paint a moderately sized room in 2 hours. How much time would it take 3 average people to paint a room that had 4 times as much area to paint?

A 8 hours   **B** 6 hours   **C** $4\dfrac{1}{3}$ hours   **D** $2\dfrac{2}{3}$ hours   **E** $1\dfrac{1}{3}$ hours

4. At a certain company, t typists can complete a company report in h hours. How many minutes would it take n typists, working at the same rate, to complete the report?

A $\dfrac{60t}{hn}$ **B** $60thn$ **C** $\dfrac{thn}{60}$ **D** $\dfrac{th}{60n}$ **E** $\dfrac{60th}{n}$

5. In a school, the ratio of large classrooms to small classrooms is 3:4, and the ratio of offices to small classrooms is 1:8. What is the ratio of large classrooms to offices?

     A 6:1   **B** 3:2   **C** 2:3   **D** 1:6

     E The answer cannot be determined from the given information.

6. During a 162-game season, after the first 27 games, a baseball team has a win-loss ratio of 5:4. If the team wins the next g games, it will have won 70% of all the games it played to that point. What is the value of g?

     A 2   **B** 10   **C** 13   **D** 15   **E** 113

7. If the ratio of two angles of a parallelogram is 7:2, what is the ratio of the other two angles of the parallelogram?

     A 1:1   **B** 2:1   **C** 3:2   **D** 5:4   **E** 7:2

8. The perimeters of two rectangular rooms are in the ratio of 3:2. If the sum of their areas is 1,040 square feet, what is the difference in their areas?

     A 25 square feet   **B** 208 square feet   **C** 400 square feet

     D 6,400 square feet   **E** 43,264 square feet

9. If 9 ounces of cereal will feed 2 adults or 3 children, then 90 ounces of cereal, eaten at the same rate, will feed 8 adults and how many children?

     A 8   **B** 12   **C** 15   **D** 18   **E** 30

10. If 30 black pencils cost x cents and 72 red pencils cost y cents, which of the following represents the cost of 5 black pencils and 12 red pencils?

     A $\dfrac{x+y}{6}$ **B** $\dfrac{x}{30}+\dfrac{y}{72}$ **C** $\dfrac{x+y}{102}$ **D** $\dfrac{72y+30x}{102}$ **E** $\dfrac{12x+5y}{360}$

11. Three circles are concentric with center D. If $AB:BC:CD =$

2:3:5, which of the following represents the ratio of the shaded area and the area of the largest circle?

A 0.25   **B** 0.35   **C** 0.39   **D** 0.40   **E** 0.64

(The figure is not necessarily drawn to scale.)

12. An inverted right circular cone is sliced parallel to its base along a plane that passes through the midpoint of the altitude. Which of the following could be the ratio of the volumes of the two solids formed?

     A 1:1   **B** 1:2   **C** 1:4   **D** 1:7   **E** 1:8

13. In an urn containing only blue and red marbles, the ratio of red marbles to blue marbles is 8:11. If the number of red marbles is doubled while the number of blue marbles is tripled, making a new ratio of red marbles to blue marbles, which of the following must be true?

     (*I*) The new ratio is greater than the original ratio.

     (*II*) The new ratio is less than $1/2$.

     (*III*) The new ratio is the same as the original ratio.

         A I only   **B** II only   **C** III only   **D** I and II only

                 E II and III only

14. If x is a positive integer and the ratio $x:y$ is 1:5, which of the following could be the value of the fraction x^2/y?

     (*I*) 1

     (*II*) $\dfrac{1}{5}$

     (*III*) 2

         A I only   **B** II only   **C** I and II only   **D** II and III only

                 E I, II, and III

Solutions to Additional Problems

Solution to Problem 1

The first task in the problem is to make sure that you are working with the same units. The choices are all meant to be *ounces*, while the problem states that you are considering two *quarts*. You have to know that there are 32 ounces in every quart of liquid. (See the reference table at the back of the book.) Two quarts, therefore, contain 64 ounces.

Let x be the multiplier. We have that there are $3x$ ounces of fruit juice and $5x$ ounces of water in the 32 ounces. The equation to solve for x is $3x + 5x = 64$, which simplifies to $8x = 64$ and $x = 8$. Therefore, there are 40 ounces of water and 24 ounces of fruit juice, choice D.

Alternative Solution to Problem 1

Using the choices, we would have the following situations if we needed a total of 64 ounces of liquid:

A 3 ounces of fruit juice and 61 ounces of water
B 6 ounces of fruit juice and 58 ounces of water
C 12 ounces of fruit juice and 52 ounces of water
D 24 ounces of fruit juice and 40 ounces of water
E 32 ounces of fruit juice and 32 ounces of water

The only choice that creates the ratio of 3:5 is choice D.

Solution to Problem 2

Let x be the multiplier. The cost of the more expensive coat would be $7x$, and the cost of the less expensive coat would be $5x$ dollars. The sum of the costs of the coats is $12x = d$ dollars and $x = {}^{d}/_{12}$. The difference of the costs of the coats is $2x$ or

$$\frac{2d}{12} = \frac{d}{6} \text{ dollars}$$

or choice B.

Alternative Solution to Problem 2

Using substitution, we can let the cost of the expensive coat be $70 and the cheaper coat be $50. Therefore, the sum of the costs $d = 120$ dollars. The difference of the costs is $20. With $d = 120$, the choices become the following:

A 10   **B** 20   **C** $\dfrac{240}{7} = 34\dfrac{2}{7}$ **D** $\dfrac{240}{5} = 48$ **E** $\dfrac{5}{7} \times 120 = 85\dfrac{5}{7}$

The answer is clearly choice B.

Solution to Problem 3

Since the problem refers to all the people as "average people," we can assume the same rate of work for all of them. (In fact, if we don't, we can't solve the problem.) The work formula is the following:

$$Rate \times time = number\ of\ jobs\ completed$$

The first sentence of the problem tells us that the rate is $1/2$ room per hour. However, with 3 people working together, the rate is $1/2 \times 3 = 3/2$ rooms per hour. The number of jobs we need completed is 4. Therefore, if we let t represent the time, the work formula gives us the equation $3/2 \, t = 4$. Solving this equation gives us $t = 4 \div 3/2 = 8/3 = 2\,2/3$ hours, choice D.

Alternative Solution to Problem 3

Using arrows we have

$$\text{people} \rightarrow \text{rooms} \rightarrow \text{hours}$$

The given information is

$$1 \rightarrow 1 \rightarrow 2$$

A room 4 times as large is the same as having 4 rooms, and it would take the 1 person 4 times as long. Therefore, we have

$$1 \rightarrow 4 \rightarrow 8$$

Now, when people work together to complete the same job, the number of people and the number of rooms have an *inverse* relationship. That is, as the number of people enlarges, the number of hours grows smaller by the same factor. Therefore, for 3 people we have the following:

$$3 \rightarrow 4 \rightarrow 8 \div 3 = 2\frac{2}{3} \text{ hours}$$

Solution to Problem 4

The information in the problem tells us that the rate of work is $1/t \div h$, or $1/th$ jobs per hour. If n people are working at the same rate, then each is doing $1/n$ jobs. The work formula, rate $\times$ time = jobs completed, with x for time gives us the following:

$$\left(\frac{1}{th}\right)x = \frac{1}{n}$$

Solving for x, we have

$$x = \frac{1}{n} \div \frac{1}{th} = \frac{1}{n} \times \frac{th}{1} = \frac{th}{n} \text{ hours, or } \frac{60th}{n} \text{ minutes}$$

which is choice E.

Alternative Solution to Problem 4

Using substitution and, then, using arrows would be the most concrete approach to solving the problem. Suppose $t = 5$, $h = 20$, and $n = 15$. The problem is that 5 typists can complete the report in 20 hours, and the task is to find the time, in minutes, that 15 typists would take.

Using arrows we have

typists $\rightarrow$ hours   in an inverse relationship

$5 \rightarrow 20$

$15 \rightarrow 20 \div 3 = 6\dfrac{2}{3}$ hours, or 400 minutes

Using $t = 5$, $h = 20$, and $n = 15$, the choices become the following:

A $\dfrac{60 \times 5}{20 \times 15} = \dfrac{300}{300} = 1$

B $60 \times 5 \times 20 \times 15 = 90,000$

C $\dfrac{5 \times 20 \times 15}{60} = \dfrac{1,500}{60} = 25$

D $\dfrac{5 \times 20}{60 \times 15} = \dfrac{100}{900} = 1/9$

E $60 \times 5 \times 20 \div 15 = 6,000 \div 15 = 400$

which is our answer.

Solution to Problem 5

Let x be the multiplier for the first ratio. We have $3x$ large classrooms and $4x$ small classrooms. We can now choose a different multiplier for the second ratio, say, y. Therefore there are y offices and $8y$ small classrooms. We are looking for the ratio of large classrooms to offices, which is represented by $3x{:}y$. Since $4x = 8y$, we have that $x = 2y$. Therefore, the ratio is as follows:

$$\frac{3x}{y} = \frac{6y}{y} = 6, \text{ or } 6{:}1$$

or choice A.

Alternative Solution A to Problem 5

Using arrows, we can set up the following:

large classrooms $\rightarrow$ small classrooms $\rightarrow$ offices

$3 \rightarrow 4$

$6 \rightarrow 8$ $\rightarrow 1$

and we immediately see that the ratio of large classrooms to offices is 6:1.

Alternative Solution B to Problem 5

Another way to solve the problem involves thinking of the ratios as fractions and multiplying them in a way that gives the desired ratio. That is:

$$\frac{\text{Large classrooms}}{\text{Small classrooms}} \times \frac{\text{small classrooms}}{\text{offices}} = \frac{\text{large classrooms}}{\text{offices}}$$

$$\frac{3}{4} \times \frac{8}{1} = \frac{24}{4} = \frac{6}{1}$$

This is a useful technique when you are asked to find specific ratios when given other ratios containing related information.

Solution to Problem 6

The ratio of 5:4 for the first 27 games, with x as a multiplier, gives us the equation $9x = 27$ and $x = 3$. Therefore, the team has won 15 games and lost 12. After playing and winning all of the next g games, the wins-to-games played ratio is the following:

$$\frac{15 + g}{27 + g}$$

Remembering that percentage is a ratio whose denominator is 100 allows us to set up the following proportion:

$$\frac{15 + g}{27 + g} = \frac{70}{100} = \frac{7}{10}$$

The cross product gives us

$$150 + 10g = 189 + 7g \quad \text{or} \quad 3g = 39 \quad \text{and} \quad g = 13$$

which is choice C.

Alternative Solution to Problem 6

Since the team's win-loss ratio is 5:4, the actual numbers could be 5 wins and 4 losses, 10 wins and 8 losses, or 15 wins and 12 losses. The last possibility must be the correct one to account for 27 games. Using each of the choices, we come up with the following new wins-to-games-played ratios:

A $\dfrac{17}{29} = 58.6\%$ **B** $\dfrac{25}{37} = 67.6\%$ **C** $\dfrac{28}{40} = 70\%$ **D** $\dfrac{30}{42} = 71.4\%$

E $\dfrac{128}{140} = 91.4\%$

Choice C matches the given information.

Solution to Problem 7

The solution is obvious if you know the fundamental properties of a parallelogram. (See the reference table in the back of the book.) A key property is that the measures of opposite angles are equal while the sum of the noncongruent angles is 180°. Since the given ratio is not 1:1, we know the ratio is referring to the noncongruent angles. Since each of the other angles is congruent to one of these, the angles must be in the same ratio. Therefore, the correct answer is E.

Alternative Solution to Problem 7

(This would only be recommended if you could not remember the key property mentioned above.) Using the choices and x as a multiplier, each choice would give a different equation based on the fact that the sum of the angles of a quadrilateral is 360°:

A $11x = 360°$ **B** $12x = 360°$ **C** $14x = 360°$ **D** $18x = 360°$
E $18x = 360°$

Choices B, D, and E give integer values of x, and so it would be preferable to try these choices before trying A and C. Choice B gives $x = 30°$. The angles of the quadrilateral would be $7x$, $2x$, $2x$, and x or 210°, 60°, 60°, and 30°. When drawing a four-sided figure with these angles, you would immediately see that it would not be the familiar shape of a parallelogram. Choice D gives us $x = 20°$ and $7x$, $2x$, $5x$, and $4x$ or 140°, 40°, 100°, and 80°. Once again, the familiar shape of a parallelogram does not appear. For choice E, $x = 20°$ and the angles are $7x$, $2x$, $7x$, and $2x$ or 140°, 40°, 140°, and 40°. These angles make it possible to create a parallelogram.

Solution to Problem 8

If the perimeters are in the ratio of 3:2, then the areas are in the ratio of 9:4. Using x as a multiplier, we have that the areas of each room are represented by $9x$ and $4x$ and the sum is $13x$. Therefore, we have $13x = 1,040$ and $x = 80$. The difference of the areas is $9x - 4x = 5x$, or 400 square feet, choice C.

Alternative Solution to Problem 8

If you do not remember that the ratio of the areas is the square of the ratio of the perimeters, you can work with the choices. However, you still need to realize that somewhere in the problem a ratio somewhat related to 3:2 must appear.

Examining the choices, you should easily realize that D and E are not possible since the difference must be less than the sum. You can use each of the remaining choices as a difference and work backward. For choice A, the difference of the areas is 25 and the sum is the given 1,040. There is still some algebra involved here. Let x be the area of the larger room and y be the area of the smaller room. You now have two simultaneous equations: $x + y = 1,040$ and $x - y = 25$. By adding the equations together, you have $2x = 1,065$ and $x = 432.5$. With this, $y = 407.5$. The ratio of the areas is not at all related to 3:2.

Performing a similar analysis with choice B, you would find that $x = 624$ and $y = 416$. This ratio is 3:2. Analyzing choice C, you would find that $x = 720$ and $y = 320$. This ratio is 9:4. You now have to make the choice between the two. Seeing the ratio 9:4 in front of you should trigger your memory on the important relationship between the ratio of the perimeters and the ratio of the areas.

Solution to Problem 9

Using algebra, we first determine the rate at which adults and children each consume cereal. The given information tells us that cereal is consumed at the rates of 4.5 ounces per adult and 3 ounces per child. Eight adults will have consumed 36 ounces of cereal, leaving 54 ounces remaining for the children. Therefore, there will be $54 \div 3 = 18$ children, choice D.

Alternative Solution to Problem 9

Using arrows, we set up the following:

$$\text{ounces of cereal} \to \text{adults} \to \text{children}$$
$$9 \to 2 \quad \to 3$$

Looking for 8 adults, we multiply each part by 4:

$$36 \to 8 \quad \to 30$$

Therefore, 8 adults require 36 ounces, and we have 54 ounces remaining. We can reuse the first line and multiply it by 6 to get the answer:

$$54 \to 12 \quad \to 18$$

Solution to Problem 10

Using arrows, we have the following:

$$\text{black pencils} \to \text{cents} \qquad \text{and} \qquad \text{red pencils} \to \text{cents}$$
$$30 \to x \qquad\qquad\qquad 72 \to y$$
$$5 \to \frac{x}{6} \qquad\qquad\qquad 12 \to \frac{y}{6}$$

Therefore, the total cost is as follows:

$$\frac{x}{6} + \frac{y}{6} = \frac{x+y}{6}$$

or choice A.

Alternative Solution to Problem 10

We can substitute numbers for x and y that simplify the problem tremendously. Using $x = 30$ cents for the cost of 30 black pencils and $y = 72$ cents for the cost of 72 red pencils, we have that each type of pencil costs 1 cent. Therefore, the cost of the 17 pencils would be 17 cents. With these values of x and y, the only choice that gives 17 cents is choice A.

Solution to Problem 11

The three-part ratio $AB{:}BC{:}CD$ has to be used to get the ratio of the radii of the three circles. Let x be the multiplier. We have $AB = 2x$, $BC = 3x$, and $CD = 5x$. Therefore, the radius of the innermost circle, CD, is $5x$, the radius of the second circle, BD, is $8x$, and the radius of the largest circle, AD, is $10x$. The area of the largest circle is $\pi(10x)^2 = 100\pi x^2$. The difference of the areas of the inner circles is as follows:

$$\pi(8x)^2 - \pi(5x)^2 = 64\pi x^2 - 25\pi x^2 = 39\pi x^2$$

The ratio is

$$\frac{39}{100} = 0.39$$

or choice C.

Alternative Solution to Problem 11

The ratio of the areas of the circles will be the ratio of the squares of the radii. From the ratio $AB{:}BC{:}CD = 2{:}3{:}5$, we can easily determine that the ratio $AD{:}BD{:}CD = 10{:}8{:}5$. Therefore, the ratio of the areas of the circles is $100{:}64{:}25$. The ratio of the difference of the areas is as follows:

$$\frac{64 - 25}{100} = \frac{39}{100} = 0.39$$

70

Solution to Problem 12

The following diagram is useful:

Let r be the radius of the base of the original cone, and let h be its height. After the cone is sliced, the smaller cone has radius $1/2\, r$ and height $1/2\, h$. The formula for the volume of a right circular cone is $1/3 \times \pi \times \text{radius}^2 \times \text{height}$. The volume of the original cone is $\pi r^2 h/3$, and the volume of the newly formed smaller cone is as follows:

$$\pi \left(\frac{1}{2} r \right)^2 \frac{\left(\frac{1}{2} h \right)}{3} = \pi r^2 \frac{h}{24}$$

The remaining solid has the volume

$$\pi r^2 \frac{h}{3} - \pi r^2 \frac{h}{24} = 7\pi r^2 \frac{h}{24}$$

Therefore, the ratio of the two solids is 1:7, choice D.

Alternative Solution to Problem 12

The ratio of the volumes of two solid objects is equal to the third power of the ratio of corresponding lengths of the solids. Since we know that the newly formed cone after the cut has lengths that are $1/2$ the size of the original cone, the volume of the cone will be $1/8$ the volume of the original. Therefore, the other piece after the cut must have a volume equal to $7/8$ the volume of the original cone. Hence, the ratio of the two new solids could be 1:7.

Solution to Problem 13

Let x be the multiplier. Therefore, the urn had $8x$ red marbles and $11x$ blue marbles to start with. After adding the new marbles, there are $16x$ red marbles and $33x$ blue marbles making the ratio $16x/33x = 16/33$. This is slightly less than $1/2$ since the numerator is slightly less than half of the denominator. (You could convert 16/33 to 0.484848 . . . to see this as well.) Clearly, the only one of the statements that is true is II, and the answer is choice B.

Alternative Solution to Problem 13

You can *eliminate choices* by carefully examining them for logical or mathematical inconsistencies. That is, choices D and E can be immediately ruled out as no two of the statements can be true at the same time. Statement III can't be true since the ratios will stay the same only if the newly added amounts of marbles are in the same ratio. Statement I is also impossible since you started with more blues than reds and the blues are being increased much more than the reds are. This will cause the fraction of red/blue to become smaller as its denominator grows faster than its numerator. By elimination, statement II is the only one that can be true.

Solution to Problem 14

If we let n be the multiplier, we have that $x = n$ and $y = 5n$. We also have that $x^2 = n^2$. The fraction we seek, in terms of n, is

$$\frac{n^2}{5n} \quad \text{or} \quad \frac{n}{5}$$

Statement I is possible if $n = 5$, statement II is possible if $n = 1$, and statement III is possible if $n = 10$. Therefore, the answer is choice E.

Alternative Solution to Problem 14

We should immediately recognize that $1/5$ is a possible value for x^2/y since $x = 1$ and $y = 5$ is the simplest possible value for x/y. Choice A is eliminated, but with choices C and D remaining, we would next check whether or not I and III can be true. We can use the given fact that $y = 5x$ and proceed to examine I and II.

For statement I, we would have to have $y = x^2$. This gives us $5x = x^2$, which is satisfied by $x = 0$ and $x = 5$. Since x must be a positive integer, we have this as a possible answer if $x = 5$ and $y = 25$.

For statement II, we would have $x^2 = 2y$ or $y = 1/2 x^2$. We would then have $5x = 1/2 x^2$, or $10x = x^2$, which is solved by $x = 0$ or $x = 10$. Again, x cannot be 0, and we would have this result if $x = 10$ and $y = 50$.

Therefore, statements I, II, and III all produce possible values for x^2/y.

72

Chapter 4

Using Lists, Patterns, and Diagrams

Many problems appearing on standardized tests are most easily solved by listing all the possibilities that can arise from the given situation. Two major areas of mathematics in which it is necessary to do this often are *probability* and *logical reasoning*. In probability, the number of these possibilities is an important part of the answer. In logical reasoning, you are often asked to draw a conclusion from specific facts concerning the classification of objects. Sorting and grouping correctly are often best achieved by listing items or arranging them in a diagram.

Recognizing patterns is another skill that is regularly tested on standardized tests. People who do this successfully are usually equally successful at working in difficult situations. Standardized tests, therefore, are constructed to measure this skill. Many of these problems involve familiar properties of numbers such as classification as odd or even, classification as prime or composite, listing factors, and listing multiples. While there is usually an arithmetic or algebraic way to solve problems involving sequences of numbers or quantities, the underlying pattern of the sequence is recognizable by listing several terms and applying some logical reasoning.

In this chapter you will see problems that include all that were mentioned above so as to give you an opportunity to develop your ability to make lists, identify patterns, and draw appropriate diagrams. Further, in this chapter, the questions will not all be multiple choice. This will provide you with practice in ascertaining the correct answer when you do not have the clues or opportunities that the choices usually provide.

Problems Involving Counting and Probability

Many word problems describe activities that when performed, can result in a variety of outcomes. You could be asked to determine how many possibilities can occur or how many of the possibilities satisfy a given condition for a specific situation. In some problems you might be asked to determine the likelihood that a specific situation will occur.

To discuss these kinds of problems, you need some vocabulary that most people use to describe the components of these types of problems. Thus, the activity is referred to as the *experiment*, and the situation being measured is called the *event*. The likelihood that the event will occur is measured as a fraction between and including 0 and 1 referred to as the *probability of the event*. To create this fraction, you have to count all the possible outcomes that can arise when performing the experiment and also count all of the favorable outcomes that satisfy the condition of the event being measured. The set of all possible outcomes is called the *sample space* of the experiment. The fraction that determines the probability of the event is the following:

$$\frac{\text{Number of favorable outcomes}}{\text{Number of outcomes in the sample space}}$$

The most accurate way to determine these counts is to actually list the elements in the sample space and identify those that are favorable. There are several ways in which these lists can be made quickly and accurately, as the next few examples will show.

Example I

Three men and two women are seated in a row of six chairs in an auditorium. If the seats at either end of the row are occupied by Mr. Jones and Mr. Smith, how many different ways can the five people be seated?

A 6   **B** 12   **C** 24   **D** 48   **E** 720

Solution I

Listing all the possibilities may at first seem overwhelming. However, the fact that two of the men are at the ends of

74

the rows reduces the problem to a much simpler situation. To see this, let us first assign letters to represent the five people. Let J represent Mr. Jones, S represent Mr. Smith, M represent the third man, X represent the first woman, and Y represent the second woman. We will also have one extra chair, which we can represent by B. Initially, we have

$$J - - - - S \quad \text{or} \quad S - - - - J$$

The problem should be clearer now. That is, *how many different ways can M, X, Y, and B be arranged in the middle four chairs?* The answer to the problem will be this number multiplied by 2 since S and J can be seated in the two different ways shown above.

The listing of M, X, Y, and B should be done systematically with each of the four written first followed by a similar listing of the others:

MXYB	XMYB	YMXB	BYMX
MXBY	XMBY	YMBX	BYXM
MYXB	XYMB	YBMX	BXMY
MYBX	XYBM	YBXM	BXYM
MBXY	XBMY	YXMB	BMXY
MBYX	XBYM	YXBM	BMYX

There are 24 such arrangements for the middle four seats and, therefore, 48 possible arrangements for all six seats, choice D.

Alternative Solution 1

There is a method to calculate the number of arrangements without listing them. This requires the use of the *counting principle*, which is used when a situation involves filling slots from a finite number of objects. The counting principle states that *the number of arrangements is the product of the number of possibilities that can occupy each of the slots.*

In this problem, the six chairs are the six slots to fill. If we start at one side of the row, we know that the first seat can be occupied by one of 2 possibilities, the second seat by one of only 4 possibilities, the third seat by one of the remaining 3, the fourth seat by one of the remaining 2, the fifth seat has only 1 possibility remaining, and the sixth seat also has only 1 possibility remaining. Therefore, the number of possible arrangements given by the counting principle is $2 \times 4 \times 3 \times 2 \times 1 \times 1 = 48$.

While the counting principle, if understood, is quicker than listing the items, it may not be as clear when answering a question of probability. The next example uses the same situation in example 1 to demonstrate this.

This example will not have choices to select from. Instead, you must compute the answer and place each value in proper order in each cell of the four-cell grid. This includes using a cell for the decimal point or the / symbol to represent the fraction bar. Usually, on the test's actual answer sheet, there are columns under each cell to bubble in the number you wrote in the cell.

Example 2

Three men and two women are seated in a row of six chairs in an auditorium. If the seats at either end of the row are occupied by Mr. Jones and Mr. Smith, what is the probability that a man is sitting in the second seat from either end?

Solution 2

From having listed all the outcomes in the sample space as we did in solution 1, you can now see that the sample space consists of 24 outcomes since they represent filling seats 2 through 5. Exactly 12 of these would have M satisfy this condition. That is, all 6 in the first column have M sitting in the second seat, and 2 from each of the other three columns have M sitting in the fifth seat. Therefore, the probability is $12/24$, or $1/2$, or 0.5.

While most tests do not require you to reduce your answer in order for it to be scored as correct, the four cells in the grid cannot handle $12/24$ since it has five characters. You must, therefore, use one of the other two forms of the answer. The ways to correctly complete the grid are as follows:

		.	5

or

	1	/	2

The next example demonstrates another way to list outcomes for counting or for determining a probability.

Example 3

In a set of 10 cards, each card has written on it a different digit from 0 to 9. If 2 cards are drawn by 2 people at the same time, what is the probability that the positive difference of the cards is 2?

Solution 3

Since you need to identify the sample space of the experiment, listing all the outcomes is an appropriate method for you to use. The way in which you make the list is the key to solving the problem quickly. For this, a square table is best with the outcomes of the pick of the first person written along the top and the pick of the second person written along the left side. In each cell you can put the positive difference of the 2 cards. Since the 2 people are picking cards at the same time, they cannot pick the same card, and you can place an X in each cell along the diagonal:

	0	**1**	**2**	**3**	**4**	**5**	**6**	**7**	**8**	**9**
0	X	1	②	3	4	5	6	7	8	9
1	1	X	1	②	3	4	5	6	7	8
2	②	1	X	1	②	3	4	5	6	7
3	3	②	1	X	1	②	3	4	5	6
4	4	3	②	1	X	1	②	3	4	5
5	5	4	3	②	1	X	1	②	3	4
6	6	5	4	3	②	1	X	1	②	3
7	7	6	5	4	3	②	1	X	1	②
8	8	7	6	5	4	3	②	1	X	1
9	9	8	7	6	5	4	3	②	1	X

You can easily determine that there are 90 possible outcomes. Those cells with a 2 are circled, and you can count that there are 16 of these. Therefore, the probability is $^{16}/_{90}$, or $^8/_{45}$:

8	/	4	5

or

.	1	7	7

Alternative Solution 3

Since any one of 10 cards can be drawn on the first pick and any one of the remaining 9 on the second pick, the counting principle tells us that there are $10 \times 9 = 90$ possible pairs of cards that can be drawn. However, since we do not care about the order of the cards, there are $90 \div 2 = 45$ possibilities. We now can list the actual pairs that produce a positive difference of 2. They are 0-2, 1-3, 2-4, 3-5, 4-6, 5-7, 6-8, and 7-9. Therefore, the probability is $^8/_{45}$.

You can see that if you do not realize that the 90 involves the specific order in which the cards are drawn, while your list of 8 does not, you would arrive at the wrong answer. To avoid this error, it is preferable to list all the outcomes.

Some problems involve determining the number of objects that satisfy one or more conditions that may overlap with one another. The next example illustrates a useful technique for grouping and sorting in such situations.

Example 4

All of the theater majors at a certain university responded to a survey in which they were questioned as to what kinds of shows that each would be comfortable performing in. Of those surveyed, 21 students selected dramas, 22 selected musicals, and 16 selected comedies. Of these, 5 selected only dramas and musicals, 3 selected only comedies and musicals, 1 selected only dramas and comedies, and 6 selected all three. What is the probability that a student, picked at random, selected only musicals?

Solution 4

The phrase "of these" in the third sentence of the problem should make it clear that these groups are not mutually

78

exclusive. That is, for example, the 6 that selected all three were counted in each of the 21 who preferred dramas, the 22 who preferred musicals, and the 16 who preferred comedies. The problem is to sort these students and display the groupings to identify those who made multiple selections. An efficient tool to accomplish this is called a *Venn diagram*, which consists of overlapping circles in which each region indicates a different one of the possible single or multiple selections. A letter is used to signify what is being represented in each circle. For this problem, we can use D for dramas, M for musicals, and C for comedies.

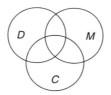

The diagram would now be filled with the numbers of students who selected all three in the innermost region, 6, and working outward to those who selected only two types of shows.

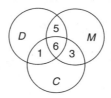

From the total numbers of students selecting each type of show, we can determine the numbers of those who selected only one type of show. That is, of those who selected dramas, $5 + 6 + 1 = 12$ are already accounted for in the dramas circle. Therefore, there are $21 - 12 = 9$ remaining who selected only dramas. There are $5 + 6 + 3 = 14$ students already in the musicals circle, leaving $22 - 14 = 8$ remaining who selected only musicals. Finally, there are $1 + 6 + 3 = 10$ students accounted for in the comedies circle, leaving $16 - 10 = 6$ remaining who selected only comedies.

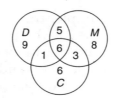

By adding the numbers in the diagram, we now know that there were 38 theater majors and exactly 8 of them would prefer only musicals. Therefore, the probability of picking a student at random who preferred to perform only in musicals is $^8/_{38}$:

8	/	3	8

or

.	2	1	1

Alternative Solution 4

Another way to arrive at the solution would be to create a table that lists the possible pairwise combinations since we know that these groups are mutually exclusive by virtue of the word *only*. We add a fourth column to represent all three areas and a fifth column to represent the total:

	Dramas	Musicals	Comedies	All	Total
Dramas	?	5	1	6	21
Musicals	5	?	3	6	22
Comedies	1	3	?	6	16

Looking at the total in each row, we see that 9 must have selected only dramas, 8 selected only musicals, and 6 selected only comedies. The total number, however, is not as clear here as it was in the Venn diagram. The total number would be found by crossing out those cells in the table in which duplication of pairs occurs (the total is irrelevant here):

	Dramas	Musicals	Comedies	All	Total
Dramas	9	5	1	6	
Musicals	8̶	8	3	6̶	
Comedies	1̶	8̶	6	6̶	

These numbers now give us the total count of 38 and the probability is $^8/_{38}$.

Example 5

In a debate on two issues among 32 people, 16 agreed with the first issue, 10 agreed with the second issue, and of these 7 agreed with both. What is the probability of selecting a person at random who did not agree with either issue?

A $\dfrac{1}{32}$ **B** $\dfrac{13}{32}$ **C** $\dfrac{3}{8}$ **D** $\dfrac{3}{16}$ **E** 0

Solution 5

We can use a Venn diagram in this problem. However, there are two major differences between this problem and example 4:

(1) There are only *two* categories that intersect, one for each issue.
(2) It is possible for a person to *not be within either category*.

For this we need a picture that contains two intersecting circles and a place to demonstrate that there is a region outside the circles. We, therefore, draw the circles within a rectangle, called the *universe*, which, in total, contains all the objects mentioned in the problem. Specifically, in this problem, we are told that our universe contains 32 people.

Starting with a 7 in the intersection of the two circles and calculating the remainder from the 16 and 10, we have the regions completed in the circles as follows:

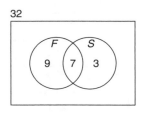

This tells us that 19 of the people agree with at least one issue. Therefore, to complete the universe, there have to be 13 people who disagreed with both issues. The probability is $^{13}/_{32}$, choice B.

Alternative Solution 5

The Venn diagram should make it clear that when there are only two categories involved in a situation, the distinct number of objects in both categories is found by the following calculation:

Number in category A + number in category B − number in both

If you know this, you can easily make the calculation that there are $16 + 10 - 7 = 19$ people who agree with at least one issue. Thus, there are $32 - 19 = 13$ who agree with neither, and the probability of selecting such a person is $^{13}/_{32}$.

Problems Involving Logical Reasoning

Tests often involve questions in which a logical conclusion has to be drawn from a good deal of given information. Listing possibilities, sorting and grouping, and making tables can often get you to the right conclusion quickly.

The following example tests your ability to recognize how information can be related in a diagram.

Example 6

In an office of a business that manufactures pencils, there are different classifications of employees. All typists are considered clerical staff, but not management. Some clerical staff are considered management, but no managers are considered executives. Using E, M, C, and T to represent executives, managers, clerical staff, and typists, respectively, which of the following best depicts this situation?

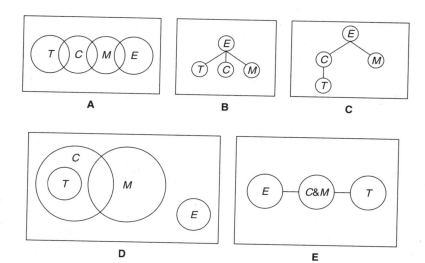

A

B

C

D

E

Solution 6

There are two types of diagrams among the five choices. Choices A and D are Venn diagrams while the other three are *flowcharts*, or *tree diagrams*. Sorting and grouping are best represented by Venn diagrams while organizational hierarchies are better represented by the other.

The first condition in the problem states "All typists are clerical staff." This, in a Venn diagram, would require that a circle representing typists must be entirely within a circle representing clerical staff. Therefore, choice D is a better representation than choice A. Among the other choices, we can eliminate choice B since it depicts typists as being entirely separated from clerical staff.

The second condition states that some clerical staff are management. This eliminates choices B and C, which depict clerical staff and management as being separated.

Now having to choose between choice D and E, you should decide which makes all the conditions clear beyond doubt. Choice E might lead someone to believe that any clerical staff that are not typists should be management. Since this is not explicitly stated, the diagram leaves doubt. Therefore, having eliminated all other choices, choice D should be the best representation of the situation. Upon checking each condition, you should be able to verify this.

Example 7

In a certain card game, all of the 52 cards are held in different players' hands at all times, but they may not necessarily have the same number of cards. Alan, Bob, Caryn, David, and Ellen are playing the game, and at one point the following situation occurs:

Bob has more cards than Alan and David combined.
Caryn has fewer cards than Ellen.
David's number is equal to the numbers of Ellen and Alan combined.

Let A represent the amount that Alan has, B represent the amount that Bob has, C represent the amount that Caryn has, D represent the amount that David has, and E represent the amount that Ellen has. Which of the following ordering of amounts from smallest to largest is possible at this point in the game?

A C, A, D, B, E **B** E, A, C, D, B **C** C, E, B, D, A
D B, E, C, A, D **E** A, C, E, D, B

Solution 7

To best understand the problem, you should attempt to list the information yourself in an organized way. In this situation, you should translate the given conditions into symbolic form using the given representations and the $<$, $>$, or $=$ symbols. It is usually helpful to use the same symbol $<$ or $>$ so that comparisons can be easily made. That is, the first condition can be symbolized by $B > A + D$, but it is more consistent with the problem and the other conditions to write it as $A + D < B$. Therefore, the three conditions become the following:

$$A + D < B$$
$$C < E$$
$$A + E = D$$

The first condition indicates that $A < B$ and $D < B$. The last condition indicates that $A < D$ and $E < D$. Therefore, any possible ordering has to have $C < E < D$ and $A < D < B$. The only one of the choices that has both of these orderings is choice E.

Alternative Solution 7

Since this is a multiple-choice problem, you can make substitutions and check the choices. The values that you substitute for the numbers of cards have to satisfy the three conditions. This can be tricky, and you may spend more time coming up with the five numbers than having been more direct in the above solution. A set of substitutions that will work is $A = 1$, $B = 6$, $C = 2$, $D = 4$, and $E = 3$. Using these numbers in the choices leaves only choice E as a true ordering of these numbers.

Problems Involving Sequences and Properties of Numbers

Sequences are generally formed by starting with a given number and following a rule that generates successive terms. For example, the sequence of positive integers is formed by starting with 1 and adding 1 to get the next term giving 1, 2, 3, 4, ..., n. Of course, most times the rule is more complex as the next examples demonstrate.

Example 8

If a term in a sequence of integers is n, the next term is $3n + 1$. If the first term in the sequence is 8, what is the largest term in the sequence less than 1,000?

Solution 8

The easiest and only sensible way to answer this question is to generate the sequence. The terms are 8, 25, 76, 247, 742, 2,227, ..., n. Clearly the answer is 742.

7	4	2

Some sequences of numbers are generated by conditional rules. That is, there may be a different rule to calculate a term if the term before it satisfies one or more conditions. A frequent condition that is checked for is whether a number is *even* or *odd*. Remember that this property applies only to the set of integers. That is, fractions are neither even nor odd. The defining condition is that *an integer is even if it has a factor of 2;*

otherwise it is odd. Therefore, negative integers can be even or odd, and 0 is considered even. You are probably well aware that even numbers are easily recognized by their ending with one of the digits 0, 2, 4, 6, or 8 while odd numbers end with one of the digits 1, 3, 5, 7, or 9.

Example 9

Each term in a sequence of positive integers is created by adding the digits of the term before it and dividing the sum by 2 if the sum is even or multiplying the sum by 2 if the sum is odd. If the first term of the sequence is 103, what is the twentieth term in the sequence?

Solution 9

The easiest way to arrive at the answer is to actually determine the terms of the sequence. Starting with 103, we have that the sum of its digits is 4. Since this is even, the next term of the sequence is 2. The sum of the digits of this term is 2, and, again, we divide by 2 to find that the next term is 1. The sum of the digits of this term is 1, and, since this is odd, we multiply by 2. You should realize that we will now alternate between 1 and 2 giving us the sequence 103, 2, 1, 2, 1, 2, 1, 2, 1, 2, ..., n.

Since every term in an even-numbered position in the sequence is 2, the twentieth term must be 2:

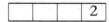

Sometimes we are not particularly interested in the sequence but rather in some property of the sequence.

Example 10

The product of 315 integers is odd. What is the least number of odd integers among these integers?

Solution 10

The key to this problem is knowing how the products of odd and even numbers become odd or even numbers themselves. Any product of integers in which there is any even integer will be an even number. This should be an obvious fact since, by definition, an even number has a factor of 2. If we form a product with an integer that has a factor of 2, then the entire product will have that factor of 2 in it. Therefore, all 315 integers in our problem must be odd.

3	1	5

It would be helpful to identify the additive and multi-plicative properties of odd and even numbers. They are:

	Addition	Multiplication
Even, even	Even	Even
Even, odd	Odd	Even
Odd, odd	Even	Odd

Other sequences that arise either in the statement or solution to a problem concern the multiples or factors of a number. The set of multiples of a number is the sequence generated by multiplying the given number by each of the numbers 1, 2, 3, 4, 5, ..., n.

Example 11

How many of the first 100 positive integers are multiples of 3 and 7?

Solution 11

The largest multiple of 7 less than 100 can be found by $100 \div 7 = 14\,{}^{2}/_{7}$, indicating that $7 \times 14 = 98$ is it. Therefore, there are exactly 14 multiples of 7 between 1 and 100, and listing them is not a tedious chore. They are:

7   14   21   28   35   42   49   56   63   70   77   84   91   98

It is easy to identify the multiples of 3 within this list. They are 21, 42, 63, and 84. The answer is 4:

			4

The next example demonstrates how the search for multiples is embedded in a more complex situation.

Example 12

Two buses begin their travels every morning at 8 a.m. at the Main Street bus stop on their routes. Bus 16 returns to this bus stop every 15 minutes while bus 5 returns every 18 minutes. By noon, how many times have the buses arrived at this bus stop at the same time?

Solution 12

The key to this problem is to examine the lists of multiples of 15 and 18 and identify the number of common multiples in both lists. While listing the multiples can be time-consuming, it can be more easily determined by creating the list in rows containing only a few multiples. We could list the numbers as follows:

$$15 \quad 30 \quad 45 \quad 60$$
$$75 \quad 90 \quad 105 \quad 120$$
$$135 \quad 150 \quad 165 \quad 180$$
$$195 \quad 210 \quad 225 \quad 240$$

This simplifies the listing of the multiples since by having stopped the first row at 60, the next row is easily found by adding 60 to the term above it and continuing in this fashion to generate more rows. We would stop here since we are considering only the 4 hours or 240 minutes from 8 a.m. to noon.

The multiples of 18 contain 90 so we might construct this list as follows:

$$18 \quad 36 \quad 54 \quad 72 \quad 90$$
$$108 \quad 126 \quad 144 \quad 162 \quad 180$$
$$198 \quad 216 \quad 234 \quad 252 \quad 270$$

The only common multiples in both lists are 90 and 180. Therefore, there are only 2 times between 8 a.m. and noon that the buses return to Main Street at the same time.

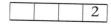

Alternative Solution 12

Another way to identify common multiples is to factor the starting numbers into a product of primes and create numbers that would have the same product of factors. That is, $15 = 3 \times 5$ and $18 = 2 \times 3 \times 3$. Therefore, the first common multiple would have to share all these factors and is $2 \times 3 \times 3 \times 5 = 90$. The common multiples would have to be multiples of 90. This list is 90, 180, 270, 360, ..., n, and only the first two fall within the 4 hours, or 240 minutes, we seek. This method is advantageous when the lists become far too long to write out.

Divisibility is another property of numbers that often appears on a standardized test. The solution above demonstrates how the factors of a number can be used to identify common multiples. Another aspect of divisibility deals with *remainders* as in the next example.

A key fact about remainders is that *When dividing by a number n, the only possible remainders are the integers 0, 1, 2, ..., n − 1*. For example, when dividing by 5, the only possible remainders are 0, 1, 2, 3, and 4. Furthermore, the remainders that occur when dividing consecutive integers by the same number will continually cycle in order. For example, if we divide the consecutive integers beginning with 1 by 5, the sequence of remainders is the following:

Integer	1	2	3	4	5	6	7	8	9	10	11	12	—
Remainder	1	2	3	4	0	1	2	3	4	0	1	2	—

Therefore, each possible remainder is the same for every fifth consecutive integer. Another way to put this is that for each multiple of 5 that leaves a remainder of 0, the next four consecutive integers will leave remainders of 1, 2, 3, and 4, respectively, when divided by 5.

Example 13

What is the smallest integer greater than 100 that leaves a remainder of 1 when divided by any of the integers 3, 4, or 5?

Solution 13

The solution to this problem can be found by applying the last idea as follows. The smallest number greater than 100 that leaves a remainder of 1 when divided by 3 is 103. Therefore, the next several numbers are possible to list easily since they will all be 3 apart:

$$103, 106, 109, 112, 115, 118, 121, 124, 127, 130, \ldots, n$$

The list of numbers greater than 100 that leave a remainder of 1 when divided by 4 will be the following:

$$101, 105, 109, 113, 117, 121, 125, 129, 133, \ldots, n$$

The list of numbers greater than 100 that leave a remainder of 1 when divided by 5 will be the following:

$$101, 106, 111, 116, 121, \ldots, n$$

The first number common to all three lists is 121:

	1	2	1

Alternative Solution 13

A quicker way to solve the problem is to locate the common multiples of 3, 4, and 5. These would all have remainders of 0 when divided by any of these numbers. The next consecutive integers would all have remainders of 1. Since 3 and 5 are prime and $4 = 2 \times 2$, the least common multiple of 3, 4, and 5 is $2 \times 2 \times 3 \times 5 = 60$. Therefore, the common multiples of 3, 4, and 5 are only the multiples of 60—namely, 60, 120, 180, 240,..., n. Therefore, the numbers 61, 121, 181, 241, 301, ..., n are the only ones that will have a remainder of 1 when divided by 3, 4, and 5.

The additional problems that follow will give you more practice with solving problems that require identifying arrangements, patterns, and groupings. Try to solve each problem in more than one way to give you additional opportunities to practice making lists, tables, and diagrams.

Additional Problems

1. How many three-letter arrangements can be formed from the letters A, D, F, G, M, O, and S if the first letter must be D, the arrangement has to include G, and no letter can be used more than once?

2. In a bookcase there are only history books, novels, and reference books. If the probability of randomly selecting a history book is $1/7$ and of randomly selecting a novel is $1/3$, what is the probability of randomly selecting a reference book?

3. Set S contains all the distinct factors of 12, 15, and 36 greater than 1. What is the probability of randomly picking two different members of the set whose product is a multiple of 10?

   A 0   **B** $\dfrac{1}{36}$ **C** $\dfrac{2}{10}$ **D** $\dfrac{4}{15}$ **E** 1

4. The winning school at a gymnastics meet involving four schools was determined by the total number of points accumulated. In each of 3 events, 5 points were awarded to the first-place team, 3 points were awarded to the second-place team, and 2 points were awarded to the third-place team. At the meet, the following occurred:

   Applegate High School finished in first place only in the first 2 events.
   Baxley High School did not finish in third place in any event.
   Calhoun High School finished in third place only in the last 2 events.
   Dixon High School earned only 5 points for having finished second in only 1 event and third in another.

   What is the probability that Baxley High School won the meet?

5. The results of a survey among teenagers indicate that $3/4$ of the males enjoy action films while $5/6$ enjoy horror movies. If $1/12$ of the

males do not enjoy either, what fraction of the male population enjoys both?

6. Every member in a community organization has to serve on at least one of three committees. There are 20 people on the social committee, 14 people on the fund-raising committee, and 10 people on the rules committee. Only 5 people are on all three committees, and 8 serve on exactly two committees. How many people are members of the organization?

7. Usually, if someone likes apples, she also likes bananas. If she doesn't like apples, then she surely likes cherries. If someone likes both dates and bananas, she will eat in the morning. Stacey likes dates, but she hasn't eaten in the morning. If Stacey's behavior demonstrates all of the above, which of the following must be true?

 A Stacey likes apples.   **B** Stacey did not eat an apple.
 C Stacey likes cherries.   **D** Stacey ate a banana.
 E Stacey does not like any fruit except dates.

8. The sum of 101 integers is even. What is the least possible number of even integers among these numbers?

 A 0   **B** 1   **C** 2   **D** 100   **E** 101

9. A candy factory makes two kinds of chocolate bars with different machines. The machine that makes chocolate bars with nuts completes a batch of 750 every 12 minutes while the machine that makes chocolate bars without nuts completes a batch of 1,000 every 10 minutes. When each machine completes a batch, its bell is rung. If the machines are both started at 7 a.m., how many chocolate bars of both kinds have been made when the bells have rung simultaneously for the fifth time?

 A 48,750   **B** 30,000   **C** 18,750   **D** 9,750   **E** 8,750

10. Let A be the set of positive factors 99 and B be the set of positive factors of 91. How many different sums are obtained when a member of set A is added to a member of set B?

11. A coach instructs the team photographer to have the camera take a picture of the action every 28 seconds. If the game lasts $2^1/_2$ hours, how many pictures will the coach have to look at?

Solutions to Additional Problems

Solution to Problem 1

The counting principle will easily solve the problem. There are three slots to fill. There is only 1 choice for the first slot. Then, there are two possibilities. Either the second slot is a G or the third slot is a G. The remaining slot in either case has 5 choices. The answer is found by the calculation

$$1 \times 1 \times 5 + 1 \times 5 \times 1 = 5 + 5 = 10$$

		1	0

Alternative Solution to Problem 1

Listing the arrangements is easy and, if done systematically, should take less than a minute. The arrangements are the following:

D G	A	D	A G
	F		F
	M		M
	O		O
	S		S

Clearly, the answer is 10.

Solution to Problem 2

The easiest solution uses the fact that the sum of all possible distinct outcomes of an experiment is 1. The sum of the two given probabilities is $1/7 + 1/3 = 10/21$. Therefore, the probability of the remaining possibility is $11/21 = 0.524$.

.	5	2	4

Alternative Solution to Problem 2

In this problem you can envision that the bookcase has three shelves, each containing a different type of book. You should realize that by finding a common denominator of the fractions, you can easily find the number of books for the first two types. Since the least common denominator is 21, the history shelf would have 3 books, and the novel shelf would have 7 books:

History	3
Novel	7
Reference	$21 - 10 = 11$

Therefore, the reference shelf would have the remaining 11 books and the probability has to be $^{11}/_{21}$.

Solution to Problem 3

The factors of 12 are 1, 2, 3, 4, 6, and 12. The factors of 15 are 1, 3, 5, and 15. The factors of 36 are 1, 2, 3, 4, 6, 9, 12, 18, and 36. Therefore, S is the set consisting of the 10 numbers 2, 3, 4, 5, 6, 9, 12, 15, 18, and 36. The key to the solution is that a multiple of 10 must have a factor of 5 along with an even number. Therefore, 5 or 15 must be one of the factors selected and another factor must be one of the remaining 6 even numbers.

Using the counting principle, we see that there are $10 \times 9 = 90$ different ordered arrangements when selecting two members of S. That is, there are 90 outcomes in the sample space. The event of having picked a multiple of 10 requires that 5 or 15 is picked either first or second and that the other number is even. Therefore, there are $2 \times 6 + 6 \times 2 = 24$ ways that this can be accomplished. The probability is $^{24}/_{90}$ or $^{4}/_{15}$, choice D.

Alternative Solution to Problem 3

Listing all the possible products in a 10×10 table is more time-consuming, but doing so will remove any doubt as to the answer. The diagonal is empty since the numbers picked must be different. This also makes it clear that the sample space consists of $100 - 10 = 90$ outcomes. If you realize that only even multiples of 5 or 15 are multiples of 10, you will have identified all 24 without having to complete the entire table.

	2	3	4	5	6	9	12	15	18	36
2	X			10				30		
3		X								
4			X	20				60		
5	10		20	X	30		60		90	180
6				30	X			90		
9						X				
12				60			X	180		
15	30		60		90		180	X	270	540
18				90				270	X	
36				180				540		X

Solution to Problem 4

Listing all the outcomes is an appropriate method for you to use in order to see the situation clearly. If we let A represent Applegate High

94

School, B represent Baxley High School, C represent Calhoun High School, and D represent Dixon High School, each of the possible outcomes for the meet can be listed in a table as shown below:

	First	Second	Third
Event I	A		
Event II	A		C
Event III			C

We can complete more of the table by analyzing the given information. Since no letter can appear in more than one cell in any row, we can deduce that Dixon must have finished in third place in event I and that Baxley must have finished in first place in event III. We now have the following:

	First	Second	Third
Event I	A		D
Event II	A		C
Event III	B		C

The possible outcomes are now determined by listing all the possibilities for second place. In completing these possibilities, remember that D must have finished second in either event II or event III. This means that either B or C finished in second place in event I. The tables now become as follows:

	First	Second	Third
Event I	A	B or C	D
Event II	A	D	C
Event III	B		C

or

	First	Second	Third
Event I	A	B or C	D
Event II	A		C
Event III	B	D	C

In the first table, only A is a possibility for the empty cell, while in the second table, only B is a possibility for the empty cell.

Therefore, there are four possible outcomes for the meet. They are listed below with the point scores for each team:

	First	Second	Third
Event I	A	B	D
Event II	A	D	C
Event III	B	A	C

$A = 13$, $B = 8$, $C = 4$, $D = 5$, and A wins.

	First	Second	Third
Event I	A	C	D
Event II	A	D	C
Event III	B	A	C

$A = 13$, $B = 5$, $C = 5$, $D = 5$, and A wins.

	First	Second	Third
Event I	A	B	D
Event II	A	B	C
Event III	B	D	C

$A = 10$, $B = 11$, $C = 4$, $D = 5$, and B wins.

	First	Second	Third
Event I	A	C	D
Event II	A	B	C
Event III	B	D	C

$A = 10$, $B = 8$, $C = 7$, $D = 5$, and A wins.

Therefore, the probability that B won the event based on the given information is $^1/_4$:

	1	/	4

or

	.	2	5

Solution to Problem 5

Using a Venn diagram with two circles within a rectangle representing all males will clarify the situation. When working with fractions, remember that the number of the entire universe is 1. Let x be the number enjoying both, and the diagram becomes the following:

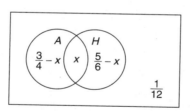

Therefore,

$$\frac{1}{12} + \frac{3}{4} - x + \frac{5}{6} - x + x = 1$$

$$\frac{20}{12} - x = 1$$

$$x = \frac{8}{12} = \frac{2}{3}$$

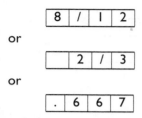

or

or

Alternative Solution to Problem 5

If the fractions bother you, you can assume that the number of male students is some multiple of 12, say, 120. Therefore, $3/4$ of $120 = 90$ enjoy action films, $5/6$ of $120 = 100$ enjoy horror films, and $1/12$ of $120 = 10$ enjoy neither. With these numbers, there are 110 students who like at least one of the two. Therefore, letting x be the number that enjoy both, we have the equation $90 + 100 - x = 110$ or $190 - x = 110$ and $x = 80$. The fraction we seek is

$$\frac{80}{120} = \frac{2}{3}$$

Solution to Problem 6

The most efficient way to solve this problem is to use a Venn diagram with three intersecting circles:

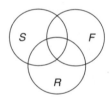

The next step is to fill in each distinct region with the number of people we can accurately determine in each group. If the number is unknown, we can use a variable. The diagram should be filled in starting in the center and moving outward.

Let x be the number of people on both the social and rules committees, let y be the number of people on both the rules and fund-raising committees, and let z be the number of people on both the fund-raising and social committees. With these, the diagram below demonstrates the numbers of people in each distinct situation.

To interpret the diagram, note that there are exactly $20 - x - z - 5$ members who serve on only the social committee, $10 - x - z - 5$ who serve on only the rules committee, and $14 - y - z - 5$ who serve on only the fund-raising committee. If we add the numbers in each distinct region,

98

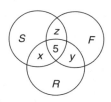

we have the total membership of the organization to be $39 - x - y - z$. The information in the problem tells us that $x + y + z = 8$. Therefore, we have $39 - (x + y + z) = 39 - 8 = 31$ members of the organization:

		3	1	

Solution to Problem 7

This problem, similar to most involving logical reasoning, is best analyzed using symbols:

Let A represent "Stacey likes apples."
Let B represent "Stacey likes bananas."
Let C represent "Stacey likes cherries."
Let D represent "Stacey likes dates."
Let E represent "Stacey eats in the morning."

The given conditions of Stacey's behavior translate to the following symbolic sentences:

(1) If A, then B.
(2) If not A, then C.
(3) If D and B, then E.
(4) D
(5) Not E.

These conditions now can be considered with regard to each being true or false based on their interrelationships with each other. Condition 4 tells us that D is true, and condition 5 tells us that E is false. We can start a table to fill in these truth values:

A	B	C	D	E
			T	F

Since E is false, the "if" clause of condition 3 must not have been met. That is, D and B are not both true. Therefore, since D is true, B must be

99

false:

A	B	C	D	E
	F		T	F

Since B is now known to be false, the if clause of condition 1 must not have been met. That is, A is false:

A	B	C	D	E
F	F		T	F

Condition 2 states that if A is false, then C must be true:

A	B	C	D	E
F	F	T	T	F

With the truth values of each individual statement having been determined, we can analyze the given choices. Clearly, choice C is the correct answer since we concluded that C is true.

Each of the other choices should be analyzed as well to be sure that they are not possible answers. Choice A would require that we had concluded that A is true. This is not the case. Choice B is not directly one of the behaviors. Stacey may have eaten an apple, even if she doesn't like them. This would be correct only if no other choice was more definite in its truth. Choice D would require that we had concluded that B is true. This is not the case. Choice E would require that we had concluded A, B, and C are all false. This is also not the case.

Solution to Problem 8

The solution requires remembering the fact that the sum of two odd integers is even. Therefore, you could have 100 of the numbers being odd integers. The remaining number would have to be even. Therefore, among the numbers, only 1 has to be an even integer.

Solution to Problem 9

The problem is most easily solved by listing the minutes that have passed when the bells have rung. The machine that makes the bars with nuts will ring at 12, 24, 36, 48, 60, 72, 84, 96, 108, 120, ..., n, minutes, and the machine that makes the plain bars will ring at 10, 20, 30, 40, 50, 60, 70, 80, 90, 100, 110, 120, ..., n, minutes. It should be clear that the first five times they will ring simultaneously are at 60, 120, 180, 240, and 300 minutes. At 300 minutes, 25 batches of chocolate bars with nuts have been produced, and 30 batches of bars without nuts have been produced. Therefore, there are $25 \times 750 + 30 \times 1,000 = 48,750$ bars combined, choice A.

Alternative Solution to Problem 9

If you realize that you are looking for common multiples of 10 and 12, then writing these numbers in their prime-factored form will get you to the least common multiple quickly. That is, $10 = 2 \times 5$ and $12 = 2 \times 2 \times 3$. Therefore, the least common multiple will be $2 \times 2 \times 3 \times 5 = 60$, and all common multiples will be multiples of 60. The fifth common multiple will be 300, and the calculations are the same as in the solution above.

Solution to Problem 10

Finding factors of a number is best accomplished by listing pairs of factors in two columns with the first ascending from 1 to the largest number less than the square root of the number. For 99, we have the following:

1	99
3	33
9	11

We clearly see that this is complete since the square root of 99 is near 10.

For 91, you may not immediately recognize factors. The key is to check through the primes. The numbers 2, 3, and 5 are not factors of 91, but 7 is. We have

1	91
7	13

Since the square root of 91 is also between 9 and 10, we have no other factors.

The sums can be found by filling in a table as follows:

	1	3	9	11	33	99
1	2	4	10	12	34	100
7	8	10	16	18	40	107
9	10	12	18	20	42	108
13	14	16	22	24	46	112

Now, going row by row, you can cross out duplicates and count those remaining to find that there are 19 different sums.

		1	9

Solution to Problem 11

The solution is easily obtained by simple arithmetic if you are clear about how to relate seconds, hours, and 28-second intervals. In $2 \frac{1}{2}$ hours, there are

$$2.5 \text{ hours} \times 60 \text{ minutes per hour} \times 60 \frac{\text{seconds}}{\text{minutes}} = 9{,}000 \text{ seconds}$$

Therefore, there are

$$9{,}000 \div 28 = 321 \frac{12}{28} \text{ intervals of 28 seconds}$$

The answer is choice B, or 321 pictures.

Alternative Solution to Problem 11

The purpose and key to making a useful list is to allow for easy computations. In this problem you need to work with multiples of 28. They are:

28 56 84 112 140 168 196 224 252 280

You can see that 280 seconds is 20 seconds short of 5 minutes. That is, 10 pictures are taken every 4 minutes 40 seconds.

Furthermore, in 2,800 seconds there are 40 minutes and 400 seconds or 40 minutes 360 seconds and 40 seconds or 46 minutes and 40 seconds. That is, 100 pictures are taken every 46 minutes and 40 seconds.

In 2.5 hours, there are 150 minutes. Listing multiples of 46 minutes 40 seconds, we have the following:

Elapsed time	Pictures taken
46 minutes 40 seconds	100
92 minutes 80 seconds	100
138 minutes 120 seconds = 140 minutes	100

During the remaining 10 minutes we see that 21 more pictures can be taken. That is,

4 minutes 40 seconds	10
8 minutes 80 seconds = 9 minutes 20 seconds	10

In the remaining 40 seconds, only one more picture can be taken. Therefore, the answer is 321.

Chapter 5

Problems Asking for a Comparison

Another kind of problem you usually don't see in the classroom or in textbooks is referred to as a *quantitative comparison*. However, this type of problem is especially popular on standardized tests. When you are being asked to make a quantitative comparison, you are usually given a small amount of information about two quantities in two columns marked A and B. You have to decide among the following choices:

A The quantity in column A is greater.
B The quantity in column B is greater.
C The quantities are equal.
D It is not possible to determine A, B, or C.

There are a variety of ways to approach these problems depending on the math topic in the question. Most questions involving arithmetic and measurement can be solved by making the computations asked for. Questions involving algebra containing variables are trickier. You will see in the examples below that many of the strategies that have been discussed in previous chapters will be useful in making the correct determination.

An important idea to keep in mind is to consider *all* possibilities. Since you are given limited information in these problems, you must not necessarily assume that your first answer is the only valid one.

Comparisons Involving Number Properties

Example I

Column A	Column B
The smallest prime integer greater than 24	The smallest even integer greater than 28

Solution I

This is an easy one! Since there is definitely a number for each of the conditions, the answer cannot be D. Furthermore, prime numbers (except for 2) are all odd, and, therefore, the numbers cannot be equal. The number in column B, the smallest even integer greater than 28, is 30. The number in column A, the smallest prime integer greater than 24, is 29. Therefore, the answer is B.

Example 2

p and q are positive integers and $p < q$:

Column A	Column B
The number of different prime factors of p	The number of different prime factors of q

Solution 2

At first glance, you might think that the larger number would have more prime factors than the smaller, and you would answer with B. If you remember to consider *all* possibilities, however, you will easily see that this is not the case.

Remember that a prime number has only one prime factor, itself. If p and q were both prime numbers, the quantities in columns A and B would be equal. On the other hand, if p were prime and q wasn't, say, $p = 5$ and $q = 6$, then q would have two prime factors (2 and 3) and the quantity in column B would be greater. Since we cannot be sure without knowing the specific values of p and q, the correct answer is D.

Comparisons Involving Calculating Amounts and Percentages

Example 3

The price of a bicycle on sale is p dollars, which is 15% less than the normal price of the bicycle:

Column A	Column B
$1.2p$	The normal price of the bicycle

Solution 3

If you really understand percentages and are comfortable with algebra, you can solve the problem in the following way. Let x be the normal price of the bicycle. Then the discounted price would be

$$p = x - 0.15x = 0.85x \qquad \text{or} \qquad \frac{85x}{100}$$

Solving for x, we have

$$x = \frac{100\,p}{85}$$

which is approximately $1.18p$, which is less than $1.2p$.

Alternative Solution 3

Substituting a number for the normal price of the bicycle will lead to the value of p. An easy number to use when working with percentages is 100. If the bicycle normally costs \$100, a 15% reduction would be \$15. Then $p = \$85$. The quantity in column A would be $1.2 \times \$85 = 102$. Therefore, column A would be greater. You could try other numbers and be confident that this is the answer.

Example 4

Column A	Column B
Percent of increase when a \$10.00 price is increased by \$1.00	Percent of increase when a \$25.00 price is increased by \$2.50

Solution 4

This problem is solved by making the computation for each column. The *percent increase* is the fraction that the increase represents compared to the original amount. For column A, the fraction is $^1/_{10}$ or 0.1 or 10%, and for column B the fraction is 2.5/25, which is also 0.1 or 10%. The quantities are equal, and the correct answer is C.

Example 5

The rate of an overseas telephone call is 85 cents for the first minute and 20 cents for each additional minute:

Column A	Column B
Cost of an 8-minute phone call	Cost of a 10-minute phone call with a 20% discount

Solution 5

The problem is solved by making the computations. For column A the cost, in cents, is $85 + 7 \times 20 = 225$. For column B the cost, in cents, is

$$80\% \text{ of } (85 + 9 \times 20) = 0.80 \times 265 = 212$$

Therefore, the quantity in A is greater, and the correct answer is A.

Comparisons Involving Sorting Data

Example 6

On tests that are scored with integers from 0 to 100, Jessica scored over 90 on her last three tests after having scored below 80 on her previous two tests:

Column A	Column B
Jessica's current average	The least score Jessica needs to have an average of 92 or higher

Solution 6

It should be obvious that we don't know exactly what Jessica's current average is, and you might think the answer is D. However, we can figure out the highest possible average that she could have based on the information, which would tell us the least she needs to score on the next test. That is, at best her five scores could be 79, 79, 100, 100, and 100. The sum of these numbers is 458, and her average would be 91.6. To have an average of 92 on six tests, the sum of the tests would have to be $6 \times 92 = 552$. Therefore, she would need at least $552 - 458 = 94$, which is greater than 91.6. Therefore, the correct answer is B.

Example 7

The daily temperatures were measured to the nearest whole degree during a week in April. The median temperature for the 7 days was 62°F, and the highest temperature was 64°F.

Column A	Column B
The mean temperature to the nearest tenth of a degree	63°F

Solution 7

For this problem, you have to understand the difference between *mean* and *median*. The *mean* is the average of all the numbers in the set under consideration. The *median* is the number that appears in the middle of the list when the numbers are arranged from smallest to largest. (If there is an even number of numbers in the list, the median is the average of the two in the middle.)

From the given information, you can deduce only that if the 7 temperatures were arranged in order from smallest to largest, there would be 3 temperatures to the left of 62°F and 2 temperatures between 62°F and 64°F. That is,

$$\_\_\_\_ 62°F \_\_\_\_ 64°F$$

It is important to consider all possibilities here. One possibility could be

$$59°F \quad 60°F \quad 61°F \quad 62°F \quad 63°F \quad 63°F \quad 64°F$$

The mean of these temperatures is 61.7°F. The question you have to ask yourself is, "Is it possible to have a greater mean?" This would happen only when the unknown 5 temperatures are the most they could be. This situation is

$$62°F \quad 62°F \quad 62°F \quad 62°F \quad 64°F \quad 64°F \quad 64°F$$

The mean temperature in this list is 62.8°F. Since column A stipulates that the mean is to be taken to the nearest tenth, the answer is B.

Alternative Solution 7

In order for the mean of the 7 temperatures to equal 63°F, the sum of the temperatures must be $63 \times 7 = 441$. Therefore, the sum of the unknown 5 temperatures has to be $441 - 62 - 64 = 315$. The maximum possible sum will occur when the 3 temperatures on the left of 62°F are all 62°F, and the two between 62°F and 64°F are both 64°F. This sum is $62 \times 3 + 64 \times 2 = 314$. Therefore, the mean must be less than 63°F.

Example 8

Of 82 students on an academic team, $1/2$ are taking Spanish, $1/6$ are taking Italian, and the rest are taking French:

Column A	Column B
Number of students taking French	Number of students not taking a language

Solution 8

You might be inclined to make the computations, but if you *read the information carefully*, you see that after the fractions, "the rest" are taking French. Therefore, the number of students not taking a language is zero, which is less than any number found for column A. The correct answer is A.

Example 9

Of 72 students, $1/2$ are passing math, $1/8$ are failing math, and the rest do not yet have their averages:

Column A	Column B
The number of students waiting for their averages	27

Solution 9

Note that $1/2 + 1/8 = 5/8$, and the rest make up $3/8$ of the class. Since $3/8$ of 72 is 27, the answer is C.

Comparisons Involving Geometry

Example 10

The area of a square is 64 square inches:

Column A	Column B
The sum of the lengths of three sides of the square	The area of a different square whose sides are one-half as large as the sides of the given square

Solution 10

If the area of the square is 64 square inches, then each side of the square is 8 inches. The sum of the lengths of three sides of the square is 24 inches. The second square would have a side length of 4 inches, and its area would be 16 square inches. The answer is clearly A.

Example 11

Petersville is 23 miles from Lottsberg, and Lottsberg is 42 miles from Denton:

Column A	Column B
The distance from Petersville to Denton	23 miles

Solution 11

At first glance, you might think that the quantity in A is greater. This would be true if the cities were in the following positions (using P, L, and D to represent the cities),

$$P\underset{23}{\rule{1.5cm}{0.4pt}}L\underset{42}{\rule{2.5cm}{0.4pt}}D$$

and they could be in this position. However, you have to always consider *all* possibilities, and there is nothing in the given information that stops the following from being possible:

$$L\underset{23}{\rule{1.5cm}{0.4pt}}P\underset{19}{\rule{1.5cm}{0.4pt}}D$$

Therefore, the correct answer is D since you can't be sure which of these two situations is the actual one.

The following additional problems will give you more practice and insight in making correct comparisons between properties. Try to solve each problem in alternative ways using the strategies presented in previous chapters to practice those skills as well. Remember to consider *all* possibilities!

Additional Problems

1. p and q are integers; $p > q$:

Column A	Column B
The sum of p^2 and q^2	The positive difference of p^2 and q^2

2. The sum of two numbers is 0, and the larger number is less than 1:

Column A	Column B
The product of the numbers	1

3. The number of tickets sold for a fund-raising event attended by adults and children was p. An adult's ticket cost $3 and a child's ticket costs $2. A total of $170 was raised by ticket sales:

Column A	Column B
$3p$	170

4. The sum of two positive numbers is 7, and the positive difference of the numbers is 4:

Column A	Column B
The positive difference of the differences of the squares of the numbers	Twice the product of the numbers

5. The sum of n consecutive integers is 0 where $n > 1$:

Column A	Column B
The average of the integers	The product of the integers

6. John can repair 7 computers in 3 hours. Kim can repair 5 computers in 2 hours:

Column A	Column B
The number of computers that John can completely repair working alone in 2 hours	The number of computers that John and Kim can completely repair working together on each machine in 1 hour

7. At a book fair where prices are in whole dollars, 24 books are bought. One-third of them cost $5.00 each, and one-fourth of them cost $7.00 each. None of the remaining books sell for more than $10.00 each:

Column A	Column B
The average cost per book	$7.00

8. The original price of an item was increased by 15%. After staying on the shelf for a month, the shopkeeper reduced the new price by 15%:

Column A	Column B
The final price of the item	The original price of the item

9. In a round-robin tournament, each of 6 players is to play g games with each of the other 5 players. The winner will be the player who accumulates the most wins:

Column A	Column B
The minimum possible number of games that, if won by any player, could make that player the winner of the tournament	$7g$

10. Two sides of a triangle are equal in length, and one of its angles measures $60°$:

Column A	Column B
The length of the third side of the triangle	The length of one of the equal sides of the triangle

11. A circle whose radius is r is inscribed in a square whose diagonal is d:

Column A	Column B
d^2	$8r^2$

12. At a summer camp, counselors are instructed to read a code off of a flagpole. On each of 5 different pennants on the flagpole is a different integer from 0 to 9, and when read from top to bottom, the 5 digits represent the code. The 0 pennant can never be on top, and the sum of the pennants must be a multiple of 9:

Column A	Column B
The number of different possible codes	10,000

Solutions to Additional Problems

Solution to Problem 1

Consider all possibilities! Integers are not necessarily positive and, if q were 0, both quantities are equal to p^2. However, the quantities would be different if p and q were different nonzero integers. The correct answer is D.

Solution to Problem 2

If the sum of two numbers is 0, then they would be opposites. That is, the numbers would be x and $-x$. The product of the numbers would be $-x^2$, which is always a negative number and is, therefore, always less than 1. (Try substituting pairs of numbers whose sum is 0 to see that this is true.) The correct choice is B. Note that the fact that the larger number is less than 1 is a superfluous piece of information.

Solution to Problem 3

The comparison can be made by performing an algebraic solution. Let x be the number of adult tickets sold, and let y be the number of children's tickets sold. The two equations that represent the given information are as follows:

$$x + y = p \quad \text{and} \quad 3x + 2y = 170$$

Multiplying the first equation by 3 gives us $3x + 3y = 3p$, and subtracting the second equation from this one give us $y = 3p - 170$. The first sentence in the given information tells us that y is at least 1. This means:

$$3p - 170 \geq 1 \quad \text{and} \quad 3p \geq 171$$

Therefore, $3p > 170$, and the correct answer is A.

Alternative Solution to Problem 3

You can substitute numbers for the amounts of each ticket sold that would raise $170, say, 50 adult tickets and 10 children's tickets. With these numbers, $p = 60$ and $3p = 180$, which is greater than 170.

You now have to make sure that this would always be the case. For this, you need to be convinced that $3p$ cannot be equal to 170 or less than 170. Now, if $3p = 170$, then p would be 56 $2/3$, which is impossible since the number of tickets sold cannot have a fractional part. The greatest multiple of 3 less than 170 is 168, and if $3p = 168$, then $p = 56$. If 56 tickets were sold, the maximum way to raise money would be to sell 55 adult tickets and 1 children's ticket (since we know that at least one child attended the event). This amount would raise only $166, not the $170 we know was raised. You can now be sure that $3p > 170$ in order to satisfy the given information.

Solution to Problem 4

Let x and y be the numbers with $x > y$. The given information tells us that $x + y = 7$ and $x - y = 4$. In column A we have the quantity $x^2 - y^2$, and in column B we have the quantity $2xy$. From algebra, we recall the familiar fact that $x^2 - y^2 = (x + y)(x - y)$. Therefore, the quantity in column A is equal to $7 \times 4 = 28$.

We can actually determine the values of x and y by adding the two equations representing the given information. This give us $2x = 11$ and $x = 5.5$. Thus y must be 1.5. The product is 8.25, and the quantity in column B has the value 16.5. Therefore, A is the correct answer.

Solution to Problem 5

The average of a set of numbers is the sum of the numbers divided by the number of numbers in the set. Therefore, the average of the numbers is $0/n = 0$.

The only way to have a set consisting of more than 1 number having a sum of 0 is to include both negative and positive numbers in the set. In the case of consecutive integers, the numbers have to surround 0—for example, $^-3 + ^-2 + ^-1 + 0 + 1 + 2 + 3$. Clearly, any such set must include 0, and, when multiplied together, the product must be 0. Therefore, the columns have equal quantities, and the correct answer is C.

Solution to Problem 6

John's rate is $^7/_3$ computers per hour. Kim's rate is $^5/_2$ computers per hour. If they work together for 1 hour, they will have repaired $^7/_3 + ^5/_2 = {}^{29}/_6 = 4\,^2/_3$ computers. Together, they can *completely* repair 4 computers. John, working alone for 2 hours, will have repaired $^{14}/_3 = 4\,^2/_3$ computers, but he will have *completely* repaired only 4 computers. The quantities are equal, and the correct answer is C.

Solution to Problem 7

The information tells us that $^1/_3$ of $24 = 8$ books that sell for $5.00 each and $^1/_4$ of $24 = 6$ books that sell for $7.00 each. If the remaining 10 books sell for $10.00 each, then the total sale is $182.00 and the average price is $7.58 per book. However, the remaining 10 books could be sold for $1.00 each and the total sale would be $92.00, producing an average of $3.83. Therefore, since the average cannot be exactly determined, the correct answer is D.

Solution to Problem 8

Let x be the price of the item. The price was first raised to $1.15x$, and, therefore, the final price was 85% of $1.15x$ or $0.85 \times 1.15x = 0.9775x$, which is less than the original price. The correct answer is B.

Alternative Solution to Problem 8

Since you are dealing with percentages, you can assume a price of $100.00. The first increase in price makes the price $115.00. Since 15% of 115 is $17.25, the final price is $115.00 − $17.25 = $97.75. You would want to try this with other values for the original price to be confident that the original price was always greater.

Solution to Problem 9

There are 15 ways in which the 6 players can play each other. That is, if the players are A, B, C, D, E, and F, then:

$$
\begin{array}{ccccc}
AB & AC & AD & AE & AF \\
   & BC & BD & BE & BF \\
   &   & CD & CE & CF \\
   &   &   & DE & DF \\
   &   &   &   & EF
\end{array}
$$

Therefore, there are $15g$ games to be played. A majority of games would be $7.5g + 1$. This number must be greater than $7g$.

Solution to Problem 10

There are two cases to consider. The first is that the $60°$ is the angle between the two equal sides of this isosceles triangle (the vertex angle). Two familiar facts in geometry are as follows:

(*1*) If two sides of a triangle are equal in length, then the angles opposite these sides are equal in measure.
(2) The sum of the angles of a triangle is $180°$.

With these facts, we see that the two remaining angles are equal and their sum is $120°$. Therefore, each is $60°$, and the triangle is an equilateral triangle in which all sides are equal in length.

The second case is that the $60°$ is a base angle. If so, then the other base angle is also $60°$, forcing the vertex angle to also be $60°$. Once again, we have an equilateral triangle. Therefore, the quantities are equal, and the correct answer is C.

Solution to Problem 11

The key word in the given information is *inscribed*. In general, a polygon is inscribed in another figure if every vertex of the polygon lies on the outer figure. For example, a regular octagon inscribed in a square would look like the figure below:

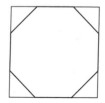

A circle inscribed in a polygon would require that the polygon have all its sides tangent to the circle. Therefore, the figure below represents the situation in this problem:

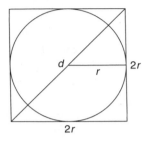

If the radius of the circle is r, then its diameter is $2r$, and this is also a side of the square. Two adjacent sides of the square along with the diagonal of the square form a right triangle with the diagonal as the hypotenuse.

The Pythagorean theorem tells us that $(2r)^2 + (2r)^2 = d^2$, which simplifies as follows:

$$4r^2 + 4r^2 = d^2 \qquad \text{or} \qquad 8r^2 = d^2$$

The quantities are equal, and the correct answer is C.

Solution to Problem 12

The three conditions to be satisfied by a code number are that each of the 5 digits is different, the first digit cannot be 0, and the code number must be a multiple of 9. In order to determine the number of possible different codes, we should create the smallest and largest possible code numbers that satisfy all three conditions.

Without considering the code number to be a multiple of 9, the smallest possible number is 10,234. When divided by 9, the number 10,234 has a remainder of 1. We can now list the multiples of 9 greater than 10,234 and stop when we have a code that has 5 different digits. That is,

$$10{,}242, \qquad 10{,}251, \qquad 10{,}260, \qquad 10{,}269, \ldots, n$$

Hence, the smallest code number that satisfies all three conditions is 10,269.

The largest possible code number without considering it to be a multiple of 9 is 98,765. This leaves a remainder of 8 when divided by 9. We can list the multiples of 9 in descending order and stop when we have a code with 5 different digits. That is,

$$98{,}757, \quad 98{,}748, \quad 98{,}739, \quad 98{,}730, \ldots, n$$

Hence, the largest code number that satisfies all three conditions is 98,730:

$$10{,}269 = 9 \times 1{,}141 \qquad \text{and} \qquad 98{,}730 = 9 \times 10{,}970$$

Therefore, eliminating the first 1,140 multiples of 9, there are $10{,}970 - 1{,}140 = 9{,}830$ multiples of 9 between and including these numbers. That is, the quantity in column A representing the number of possible codes is 9,830. This is less than 10,000, and the correct answer is B.

Chapter 6

Additional Problems for Practice

The problems that follow are similar to those found in the previous chapters. As you have seen throughout the book, there are usually several ways in which the problems can be solved. Remember, to maximize your score on standardized tests, you have to find the correct solution to the problems in the least amount of time. To accomplish this, employing one of the strategies that have been demonstrated would be preferable to attempting a traditional solution.

Problems Similar to Those in Chapter I

The solutions to problems 1 through 8 can be found by *working backward from the choices* as an alternative to working through a traditional approach.

Problem I

A pharmaceutical company manufactures three grades of aspirin—extra strength, regular strength, and mild. In a certain month, the combined sales of all three were 420,000 bottles of aspirin. If the combined number of bottles of extra-strength and mild aspirin sold was one-half of the number of bottles of regular-strength aspirin sold, what was the number of bottles of regular strength sold?

A 140,000   **B** 180,000   **C** 210,000   **D** 280,000   **E** 300,000

Problem 2

The average of four different positive numbers is 18. If one of the numbers is 28, then which of the following could be one of the remaining three numbers?

A 41   **B** 42   **C** 43   **D** 44   **E** 45

Problem 3

The product of two integers is 36, and their ratio is 4. Which of the following could be the smaller of the numbers?

A -18 **B** -12 **C** 2   **D** 3   **E** 4

Problem 4

When planning a party, Chad assumed that food would cost $150 and that paper to make invitations on his computer would cost $25. What fraction of his $210 budget was left to buy other things he needed for the party?

A $\dfrac{1}{7}$ **B** $\dfrac{2}{7}$ **C** $\dfrac{1}{6}$ **D** $\dfrac{1}{5}$ **E** $\dfrac{1}{3}$

Problem 5

Two consecutive angles of a parallelogram measure $a°$ and $b°$. If $6a - 2b < 10$, which of the following could be the measure of the larger of the two angles?

A 75   **B** 90   **C** 100   **D** 120   **E** 135

Problem 6

Two sides of a right triangle measure 15 centimeters and 17 centimeters. Which of the following statements could be true about the length of the third side?

(I) The third side is between 4 and 7 centimeters.
(II) The third side is between 20 and 23 centimeters.
(III) The third side is less than 10 centimeters.

A I only   **B** II only   **C** III only   **D** II and III only   **E** I, II, and III

Problem 7

Matt, Brad, and Katie decided that Brad and Katie would purchase CDs for their car trip and Matt would pay for the gas. Matt and Brad have $21.00 together, Matt and Katie have $36.00 together, and Brad and Katie together have one-half of the total money of all three. How much money can Brad and Katie spend to buy CDs?

A $19.00   **B** $21.50   **C** $28.50   **D** $43.00   **E** $57.00

Problem 8

p and q are different positive integers, and p is prime number less than 12. If the product $12pq$ is the cube of an integer, which of the following could not be a value of q?

A 2×3 **B** 3^2 **C** $2 \times 3 \times 5^2$ **D** $2 \times 3^2 \times 5^2$ **E** $2 \times 3^2 \times 7^2$

Problems Similar to Those in Chapter 2

The solutions to problems 9 through 16 can be found by *substituting numbers for the variables* in the problems and creating less complicated situations to solve.

Problem 9

If x is a prime number between 60 and 70, which of the following has the least value?

A $x - 30$ **B** $30 - x$ **C** $\dfrac{30}{x}$ **D** $\dfrac{x}{30}$ **E** $\dfrac{x}{x - 30}$

Problem 10

If a and b are two positive numbers such that their sum is greater than 15, which of the following is always true?

A $ab < 15$ **B** $a^2 + b^2 > 225$ **C** $a^2 + b^2 < 225$
D $(a + b)^2 = 225$ **E** $2ab = 0$

Problem 11

If p and q are unequal nonzero numbers, which of the following is the reciprocal of the difference of their reciprocals?

A $\dfrac{1}{p} - \dfrac{1}{q}$ **B** $p - q$ **C** $\dfrac{1}{p - q}$ **D** $\dfrac{p - q}{pq}$ **E** $\dfrac{pq}{p - q}$

Problem 12

Two people working separately can complete the same job in m and n hours, respectively. How many jobs can the men finish working together for h hours?

A $\dfrac{hmn}{m + n}$ **B** $\dfrac{mn}{h(m + n)}$ **C** $\dfrac{h(m + n)}{mn}$ **D** $\dfrac{m}{h} + \dfrac{n}{h}$ **E** hmn

Problem 13

A card is selected from a standard deck of 52 cards after c cards are discarded. If no picture cards were discarded, what is the probability that the selected card is a king?

A $\dfrac{1}{c}$ **B** $\dfrac{4}{c}$ **C** $\dfrac{1}{52-c}$ **D** $\dfrac{4}{52-c}$ **E** $\dfrac{4c}{52}$

Problem 14

Isabel was w years old y years ago. Her husband, who is older, will be v years old x years from now. What is the difference in their ages?

A $v-w-x-y$ **B** $v-w-x+y$ **C** $-v-w+x+y$
D $v+w-x-y$ **E** $v-w+x+y$

Problem 15

The area of a square is three times as large as the area of a triangle. If a side of the square is s inches and the base of the triangle is b inches, the number of inches in the height of the triangle is which of the following?

A $\dfrac{s^2}{3}$ **B** $\dfrac{3b}{s^2}$ **C** $\dfrac{3s^2}{2b}$ **D** $\dfrac{s^2}{3b}$ **E** $\dfrac{2s^2}{3b}$

Problem 16

Two radii of a circle are perpendicular, and a chord is drawn between the points where the radii meet the circle. If the radius of the circle is r, the area of the *larger* region between the chord and the circle is which of the following?

A $r^2\left(\pi-\dfrac{1}{2}\right)$ **B** $r^2\left(\dfrac{3}{4}\pi+\dfrac{1}{2}\right)$ **C** $r^2\left(\dfrac{1}{4}\pi-\dfrac{1}{2}\right)$

D $r^2\left(\dfrac{1}{4}\pi+\dfrac{1}{2}\right)$ **E** $r^2\left(\dfrac{3}{4}\pi-\dfrac{1}{2}\right)$

Problems Similar to Those in Chapter 3

The solutions to problems 17 through 24 can be found by *applying simple ratios and proportions.*

Problem 17

In a draw there are 30 pairs of socks consisting only of black and brown socks. If the ratio of black socks to brown socks is $3:2$, how many pairs of brown socks must be purchased to have an equal amount of both?

A 1 **B** 6 **C** 12 **D** 18 **E** 30

Problem 18

A math teacher can grade 25 exams in 30 minutes. If he usually rests for 10 minutes after grading 40 exams, how many *hours* will it take for him to grade 120 exams?

A $2\frac{11}{15}$ **B** $2\frac{2}{5}$ **C** $2\frac{9}{10}$ **D** $2\frac{17}{30}$ **E** $1\frac{2}{3}$

Problem 19

A faulty clock loses $1\frac{1}{4}$ minutes every 6 hours. If the clock is reset at noon on Monday to show the actual time, what time, to the nearest minute, will it display on Tuesday at 8 a.m. before it is reset again?

A 7:52   **B** 7:54   **C** 7:56   **D** 7:58   **E** 8:04

Problem 20

A multiple-choice test containing fewer than 100 questions is such that each question has 5 choices from which the test taker is to select the correct answer, A, B, C, D, and E. The company that creates this test is interested in the way in which students respond to the test and examines each paper to find the distribution of selected choices. On one paper the ratio of selected choices $A : B : C : D : E$ is $1 : 2 : 5 : 3 : 1$, and on another paper the ratio is $2 : 3 : 3 : 1 : 1$. Which of the following could be the number of questions on the test?

A 24   **B** 30   **C** 48   **D** 60   **E** 90

Problem 21

The sides of a quadrilateral are in the ratio of $2 : 3 : 4 : 5$. A similar figure is constructed whose perimeter is 175 centimeters. If each side of the new figure is $2\frac{1}{2}$ times as large as the original, what is the length of the largest side of the original quadrilateral?

A 2 centimeters   **B** 5 centimeters   **C** 12.5 centimeters   **D** 14 centimeters   **E** 25 centimeters

Problem 22

The ratio of the areas of two equilateral triangles is $16 : 9$. If the perimeter of the smaller triangle is 63 inches, how much larger is a side of the larger triangle than a side of the smaller triangle?

A 1 inch   **B** 3 inches   **C** 4 inches   **D** 7 inches   **E** 21 inches

Problem 23

After playing the first 20 games of the season, a team's winning percentage is 65%. If the win-loss ratio for the next 20 games is $3 : 2$, what will its new winning percentage be?

A 70% **B** 65% **C** 62.5% **D** 60% **E** 50%

Problem 24

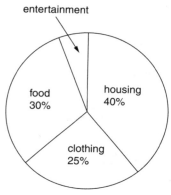

entertainment

food
30%

housing
40%

clothing
25%

Household Budget

The graph above represents the allocation of finances for a household budget. If the cost of housing decreased by 10% and the amount saved was applied to entertainment, then the increase in entertainment would be what percent of the previous allocation for entertainment?

A 4% **B** 9% **C** 50% **D** 80% **E** 100%

Problems Similar to Those in Chapter 4

The solutions to problems 25 through 32 can be found by *creating lists, drawing diagrams, and looking for patterns.*

Problem 25

If the ratio of two integers is equal to $5/9$, then their sum could equal all except which one of the following?

A −98 **B** −14 **C** 28 **D** 32 **E** 56

Problem 26

Bill has some dollar bills, and the number of Bob's dollar bills is one-half of five times Bill's amount. How many possible different amounts of dollar bills can Bill have if Bob has fewer than 25 dollar bills?

A 4 **B** 5 **C** 12 **D** 24 **E** 25

Problem 27

Five people each selected a different card at random from a standard 52-card deck. None of the cards were picture cards, and the following situations regarding the numerical values of the cards are true:

Allison's number was less than Bob's number.
Bob's number was less than Caryn's number.
Neither Danny nor Ellen has the lowest card.
Caryn has the highest number.
Danny's number is the average of Allison's and Caryn's numbers.

Which of the following ordering of people with numbers from smallest to largest are possible based on the information above?

(*I*) Allison, Bob, Danny, Ellen, Caryn
(*II*) Bob, Danny, Allison, Ellen, Caryn
(*III*) Allison, Bob, Ellen, Danny, Caryn

A I only   **B** I and II   **C** II and III   **D** I and III   **E** I, II, and III

(Questions 28 through 32 are not multiple choice, and the answer should fit into the four-character grid including the division symbol, /, or a decimal point.)

Problem 28

The first two terms of a sequence are 5 and 9. Each successive term is found by taking the average of all the preceding terms. What is the tenth term in the sequence?

Problem 29

A school uses inventory codes that are 6 characters long. The only letters used in the code are A, B, and C while the other characters are single digits from 0 to 9. The first 3 characters are different letters in any order, and the last 3 characters are different numbers in ascending order whose sum is 12. How many different ID codes are possible?

Problem 30

A survey of computer usage asked the respondents to indicate which of three applications they used frequently. The survey revealed that 80% use the Internet, 40% use word processing, and 70% play games. Further analysis revealed that 40% only play games and use the Internet, 10% only use word processing and the Internet and, 20% only play games and use word processing. If 20% only use the Internet, what percentage of the respondents indicated that they did all three?

Problem 31

Two identical six-sided cubes have a different digit from 0 through 9 written on each side, and only one of the digits is odd. What is the probability that when the dice are rolled together, the sum of the face values is odd?

Problem 32

For an integer N, $*N$ is the number $(N + 1)^2 - (N - 1)^2$. The first term of a sequence is 1, and each successive term is the $*$ value of the term before it. For example, the second term is $*1 = 4$, and the next term is the value of $*4$. What is the seventh term in the sequence?

Problems Similar to Those in Chapter 5

Problems 33 through 40 are *quantitative comparisons* taken from all areas.

Problem 33

Of the 50 students in the school play, 15 had speaking parts. Of these 15, $^2/_3$ were girls. The number of girls with speaking parts in the play is $^1/_2$ the number of boys in the play:

Column A	**Column B**
The number of girls in the play	The number of boys in the play

Problem 34

A square and a circle intersect in a plane:

Column A	**Column B**
The greatest number of points that the square and the circle have in common	5

Problem 35

Kate and Matt have $30 together, and Kate and Brad have $15 together. Together, all three have $32:

Column A	**Column B**
The amount that Matt and Brad have together	Twice the amount that Kate has alone

Problem 36

A side of a regular hexagon is equal to the radius of a circle:

Column A
The area of the hexagon

Column B
The area of the circle

Problem 37

Josh and Jennifer start at the same time and together paint their apartment in 3 hours. Jennifer always rests for 5 minutes after painting for 20 minutes, and Josh always rests for 5 minutes after painting for 30 minutes:

Column A
One-half the number of minutes they work together

Column B
The number of minutes they don't work together

Problem 38

Danny's grade on his last math test was a two-digit number divisible by 2, 3, and 7. Gayle's grade on the same test was a two-digit number divisible by 2, 3, and 5:

Column A
Danny's grade on the test

Column B
Gayle's grade on the test

Problem 39

For any number x, $[x]$ is defined to be the greatest integer less than or equal to x. p and q are numbers whose sum is negative and whose product is positive, and neither is an integer:

Column A
$[p] + [q]$

Column B
$p + q$

Problem 40

Out of 300 employees in a factory, $2/3$ make $12.00 per hour, $1/4$ make $13.50 per hour, and the rest make $14.50 per hour. Of the lowest paid employees, $1/4$ work overtime on a regular basis, which pays $1\frac{1}{2}$ times the hourly wage. No other employees work overtime:

Column A
The number of employees making more than $12.00 per hour

Column B
The number of employees making the least hourly wage

Answers to Problems

1. D
2. A
3. B
4. C
5. E
6. D
7. A
8. C
9. B
10. C
11. E
12. C
13. D
14. A

15. E
16. B
17. B
18. A
19. C
20. D
21. E
22. D
23. C
24. D
25. D
26. A
27. D
28. | | | | 7 |

29. | | | 6 | 0 |
30. | | | 1 | 0 |
31. | 5 | / | 1 | 8 |
32. | 4 | 0 | 9 | 6 |
33. A
34. A
35. B
36. B
37. A
38. D
39. B
40. C

Table of Common Measurements and Conversions

Length or Distance in Linear Units

Straight-line distance between two points: $\overline{\text{1 unit}}$

English system		Metric system		Conversions between systems	
1 inch (in)	Basic unit	1 millimeter (mm)	1/1000 m	1 in	2.54 cm
1 foot (ft)	12 in	1 centimeter (cm)	1/100 m	39 in	1 m
1 yard (yd)	36 in	1 decimeter (dm)	1/10 m	1 mi	1.61 km
1 mile (mi)	5,280 ft	1 meter (m)	Basic unit		
		1 kilometer (km)	1,000 m		

Area or Space in Square Units

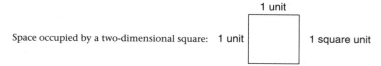

Space occupied by a two-dimensional square: 1 unit

1 unit

1 square unit

English system		Metric system		Conversions between systems	
1 square inch (in^2)	Basic unit	1 square meter (m^2)	Basic unit	10.8 ft^2	1 m^2
1 square foot (ft^2)	144 in^2	1 hectare (ha)	10,000 m^2	2.47 acres	1 ha
1 square yard (yd^2)	9 ft^2				
1 acre	43,560 ft^2				

Volume or Capacity in Cubic Units

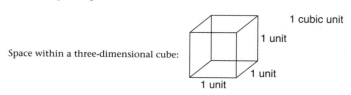

Space within a three-dimensional cube:

1 cubic unit

1 unit

1 unit

1 unit

English system		Metric system		Conversions between systems	
1 ounce (oz)*	Basic unit	1 millilter (mL)	1 cm³or cc	1.06 qt	1 L
1 cup	4 oz*	1 liter (L)	1,000 mL or 1 dm³		
1 pint (pt)	2 cups = 8 oz				
1 quart (qt)	2 pt = 32 oz	1 kiloliter (kL)	1,000 L or 1 m³		
1 gallon (gal)	4 qt or 277.42 in³				

Weight or Mass

English system		Metric system		Conversions between systems	
1 ounce (oz)†	Basic unit	1 gram (g)	Basic unit	2.2 lb	1 kg
1 pound (lb)	16 oz†	1 kilogram (kg)‡	1,000 g		
1 ton	2,000 lb	1 metric ton (t)	1,000 kg		

*These are referred to as *fluid ounces*.
†These are referred to as *dry ounces*.
‡One kilogram is equal to the weight of one liter of water.

Table of Common Formulas Seen in Word Problems

Measuring Length

Perimeter of a rectangle $= 2 \times \text{length} + 2 \times \text{width}$
or $2 \times \text{base} + 2 \times \text{height}$:

Width or height

Length or base

Circumference of a circle $= 2 \times \pi \times \text{radius}$ (π is approximately 3.14)
Distance traveled $=$ average rate $\times$ time traveled

Measuring Area

Area of a rectangle $=$ length $\times$ width or base $\times$ height

Area of a square $=$ side^2:

Side

Area of a parallelogram $=$ base $\times$ height:

Height

Base

Area of a triangle $= {}^1/_2 \times$ base $\times$ height:

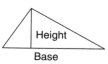

Height

Base

Area of an equilateral triangle $= \sqrt{3}/4 \times$ side^2:

Side

Area of a trapezoid = average of the bases
    × height or $\frac{1}{2}$ × (lower base + upper base)× height:

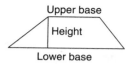

Area of a circle = π × radius^2 (π is approximately 3.14):

Surface area of a three-dimensional sphere = $4 \times \pi \times$ radius^2:

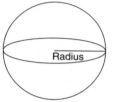

Measuring Volume

Volume of a rectangular solid = length × width
    × height or area of the base × height:

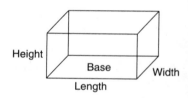

Volume of a cube = side^3:

Volume of a sphere = $\frac{4}{3}$ × radius^3

Volume of a circular cylinder = π × radius^2× height:

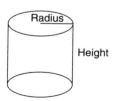

Volume of a circular cone = $\frac{1}{3} \times \pi \times radius^2 \times$ height:

Money-Related Formulas

Total dollar amount of item = quantity of item × dollar amount of one item
Total cost = price + tax rate × price
Discounted price = original price − discount × original price
Interest earned = principal × rate of interest × time invested
Current amount = principal + interest earned

Table of Fundamental Geometric and Trigonometric Relationships

Definitions

Congruent	Equal in measure and form.
Similar	Proportional in measure, but identical in form.
Supplementary angles	Two angles whose sum is 180°.
Complementary angles	Two angles whose sum is 90°.
Adjacent angles	Two angles that share a common side and vertex.
Vertical angles	Two nonadjacent angles formed by intersecting lines.
Parallel lines	Two lines that are always the same distance apart and, therefore, will never intersect.
Regular polygon	A closed figure in which all sides and interior angles are congruent.
Bisector	A line or line segment that divides a side or angle into two congruent sides or angles.
Altitude	A line segment from a vertex of a triangle perpendicular to the opposite side. Its length is the *height* of the triangle when the opposite side is considered to be its *base*.

Median   A line segment from a vertex of a
                  triangle drawn to the midpoint of
                  the opposite side, therefore,
                  bisecting the opposite side.

Relationships between Lines and Angles

(1) Vertical angles formed by intersecting lines are
congruent.
(2) Adjacent angles forming a straight line are supplementary.
(3) Adjacent angles forming a right angle are complementary.
(4) Perpendicular lines form right angles.
(5) A transversal that intersects a pair of parallel lines forms:
    a. Congruent angles in corresponding positions on each
    parallel line
    b. Congruent angles interior to both parallel lines and
    on alternate sides of the transversal
    c. Supplementary angles interior to both parallel lines
    and on the same side of the transversal

Relationships Involving Triangles

(6) The sum of any two sides of a triangle must be greater
than the remaining side.
(7) The sum of the angles of a triangle is $180°$.
(8) An exterior angle is equal to the sum of the two
nonadjacent interior angles.
(9) Two triangles are similar if they have the same three
angles.
(10) The ratios of corresponding sides of similar triangles are
equal.
(11) A median divides a triangle into two triangles of equal
area.
(12) In an isosceles triangle, two sides are congruent and
their opposite angles are congruent.
(13) In an isosceles triangle, the altitude drawn to the
noncongruent side is also an angle bisector and a median.
(14) In an equilateral triangle, all sides are congruent and
each angle measures $60°$.

Relationships Involving Right Triangles

(15) The Pythagorean theorem states that the square of the hypotenuse is equal to the sum of the squares of the legs.

(16) The square of the length of an altitude drawn to the hypotenuse of a right triangle is equal to the product of the segments it creates on the hypotenuse.

(17) The trigonometic ratios are:

a. *Sine* of an angle $= \dfrac{\text{length of opposite side}}{\text{length of hypotenuse}}$

b. *Cosine* of an angle $= \dfrac{\text{length of adjacent side}}{\text{length of hypotenuse}}$

c. *Tangent* of an angle $= \dfrac{\text{length of opposite side}}{\text{length of adjacent side}}$

Relationships Involving Quadrilaterals

(18) The sum of the angles of any quadrilateral is 360°.

(19) Opposite sides of a parallelogram are parallel and congruent.

(20) Opposite angles of a parallelogram are congruent.

(21) Consecutive angles of a parallelogram are supplementary.

(22) The diagonals of a parallelogram bisect each other.

(23) A rectangle, a rhombus, and a square are all parallelograms and have relationships 18 through 22.

(24) A rectangle has four right angles, and its diagonals are congruent.

(25) A rhombus has four congruent sides, and its diagonals are perpendicular.

(26) A square is both a rectangle and a rhombus and has relationships 24 and 25.

(27) A trapezoid has only one pair of parallel sides.

Relationships Involving Regular Polygons

(28) The sum of the interior angles of an *n*-sided regular polygon is $180 \times (n - 2)$.

(29) The measure of each exterior angle of an *n*-sided regular polygon is $360 \div n$, and the interior angle is its supplement.

(30) The segments are drawn from the center of an *n*-sided regular polygon to the vertices from *n* congruent isosceles triangles.

(31) The ratio of the areas of similar polygons is equal to the square of the ratio of corresponding sides or of the perimeters.

Relationships Involving Circles

(32) All radii of the same circle or congruent circles are congruent.

(33) The diameter is the longest line segment with endpoints on the circle.

(34) A radius meets a tangent to a circle at a right angle.

(35) Opposite angles of a quadrilateral inscribed in a circle are supplementary.

A Brief Review of Fractions

The information below will refresh your memory about some of the most important facts about fractions.

Working with Fractions

A *fraction* is a part of a whole expressed with a denominator [the lower (below the bar) number], which tells you how the whole is divided, and a numerator [the upper (above the bar) number], which tells you how many parts you have. For instance, the fraction $3/10$ indicates that the whole was divided into 10 equal parts and that you are working with 3 of those parts.

The *bar* is really a division bar, which means that the fraction is equal to $3 \div 10 = 0.3$, its *decimal equivalent*. Multiplying a decimal by 100 gives its *percentage equivalent*. Therefore, the fraction is also equivalent to 30%.

Example. $7/9 = 7 \div 9 = 0.777\ldots$, a repeating decimal. This is also equivalent to $77.777\ldots$ or $77 7/9\%$.

Working with Mixed Numbers and Improper Fractions

A *mixed number* consists of a whole number and a fraction. Usually this has to be converted to a single fraction. The whole number can be written as a fraction by finding how many of the same parts are in the whole amounts.

Example. $1 2/5$ means that you have one whole and two-fifths of a whole. There are five-fifths in the whole, and, therefore, $^7/_5$ alltogether. That is,

$$1\frac{2}{5} = \frac{5}{5} + \frac{2}{5} = \frac{7}{5}$$

When the numerator is greater than or equal to the denominator, we say that we have an *improper fraction*.

Example. $3^2/_3 = {}^9/_3 + {}^2/_3 = {}^{11}/_3$ is an improper fraction.

A quick way to convert a mixed number to an improper fraction is to multiply the whole number by the denominator of the fraction and add it to the numerator of the fraction. This tells you the total number of parts you are working with. For example,

$$7\frac{4}{9} = \frac{7 \times 9 + 4}{9} = \frac{67}{9}$$

$$5\frac{1}{8} = \frac{5 \times 8 + 1}{8} = \frac{41}{8}$$

Multiplying Fractions

The product of two fractions is found by multiplying the numerators by each other and the denominators by each other:

$$\frac{2}{5} \times \frac{8}{7} = \frac{2 \times 8}{5 \times 7} = \frac{16}{35}$$

$$\frac{4}{15} \times 2\frac{3}{5} = \frac{4}{15} \times \frac{13}{5} = \frac{4 \times 13}{15 \times 5} = \frac{52}{75}$$

Reducing Fractions

Equivalent fractions are numbers that have the same value. A fraction can be rewritten as an equivalent fraction by *reducing* it. Reducing requires finding the *greatest common factor* of the numerator and denominator, as shown in the following examples:

$$\frac{24}{36} = \frac{2 \times 12}{3 \times 12} = \frac{2}{3} \times \frac{12}{12} = \frac{2}{3} \times 1 = \frac{2}{3}$$

$$\frac{42}{54} = \frac{7 \times 6}{9 \times 6} = \frac{7}{9} \times \frac{6}{6} = \frac{7}{9} \times 1 = \frac{7}{9}$$

Some people say that you are "canceling common factors." *You really aren't canceling anything,* but instead you are using the fact that a fraction whose numerator and denominator are the same is equal to 1 and any number times 1 is itself.

This method is also helpful when you multiply fractions with large numbers. You can actually reduce the product before multiplying, as shown in the following examples:

$$\frac{25}{27} \times \frac{9}{10} = \frac{5 \times 5}{3 \times 9} \times \frac{1 \times 9}{2 \times 5} = \frac{5 \times 1 \times 5 \times 9}{3 \times 2 \times 5 \times 9} = \frac{5}{6} \times \frac{45}{45}$$
$$= \frac{5}{6} \times 1 = \frac{5}{6}$$
$$\frac{3}{7} \times \frac{7}{3} = \frac{3 \times 7}{7 \times 3} = \frac{21}{21} = 1$$

Note that the fractions $3/7$ and $7/3$ are *reciprocals*. The product of reciprocals is always 1.

Dividing Fractions

To find the quotient of two fractions, you do a strange thing. You turn the problem into a multiplication problem and multiply the first fraction by the *reciprocal* of the second fraction. In other words, dividing a number by a fraction is the same as multiplying the number by the reciprocal of that fraction, as shown in the following example:

$$\frac{8}{9} \div \frac{1}{9} = \frac{8}{9} \times \frac{9}{1} = \frac{72}{9} = 8$$

The answer makes sense if you think of the problem in the following terms: "How many times will $1/9$ go into $8/9$?" For example,

$$\frac{14}{15} \div \frac{2}{5} = \frac{14}{15} \times \frac{5}{2} = \frac{70}{30} = \frac{7}{3}$$

You can check this as follows:

$$\frac{7}{3} \times \frac{2}{5} = \frac{14}{15}$$

Adding and Subtracting Fractions

This is usually the hardest arithmetic of fractions for many people. To do it, you have to remember to be working in the same kind of parts of the whole. In other words, you need a *common denominator*. This requires changing each fraction into equivalent fractions with the same denominator. This can be done by multiplying the numerator and denominator of the fractions by the same number (the opposite of reducing). When you are working with the same kind of parts, you can find the total amount of parts you have by adding or subtracting the numerators.

The easiest common denominator is the product of the denominators. Consider the following examples:

$$\frac{4}{7} + \frac{2}{3} = \frac{4 \times 3}{7 \times 3} + \frac{2 \times 7}{3 \times 7} = \frac{12}{21} + \frac{14}{21} = \frac{26}{21}$$

or as a mixed number: $1\frac{5}{21}$

$$\frac{3}{5} - \frac{2}{9} = \frac{3 \times 9}{5 \times 9} - \frac{2 \times 5}{9 \times 5} = \frac{27}{45} - \frac{10}{45} = \frac{17}{45}$$

Sometimes you can spot a common denominator quickly and work with only one of the fractions, as in the following example:

$$\frac{7}{36} + \frac{5}{12} = \frac{7}{36} + \frac{5 \times 3}{12 \times 3} = \frac{7}{36} + \frac{15}{36} = \frac{22}{36}$$

or, when reduced: $\frac{11}{18}$

Most calculators allow you to do arithmetic with fractions. Make sure that you know how your specific calculator requires you to enter fractions into the display. *Entering fractions incorrectly can cause costly mistakes when solving problems.*

A Brief Review
of Signed Numbers

The information below will refresh your memory about some of the most important facts about signed numbers.

In the set of all numbers, some are negative and some are positive. These numbers are useful for representing opposite situations around zero such as in temperature, profit and loss, east and west of a specific point, up and down from a specific height, and many more. They are best represented on the number line:

```
<----|-----|-----|-----|-----|-----|-----|-----|-----|-----|-----|----->
    -5   -4   -3   -2   -1    0   +1   +2   +3   +4   +5
```

Fractions and decimals can also be signed. Examples are $-1/2$, $-5/2$, and $+3/5$. Pairs of numbers such as -3 and $+3$ are called *opposites* because they are on opposite sides of 0 and just as far.

Key Fact: Zero has no sign.

Key Fact: Any negative number is less than 0 or any positive number.

The set of numbers $\{\ldots, -4, -3, -2, -1, 0, +1, +2, +3, +4, \ldots\}$ that does not contain fractions is called the *set of integers.*

Absolute Value

The *absolute value* of a signed number represents its distance from zero without considering its direction. It is denoted by vertical bars around the number.

Example. $|-5| = 5$ because -5 is five units from zero and $|+7| = 7$ because $+7$ is seven units from zero.

Key Fact: *The absolute value of a negative number is its opposite.*

Example

$$|-2| = 2$$

Key Fact: *The absolute value of 0 or any positive number is itself.*

Example

$$|0| = 0 \qquad |+2| = 2$$

Adding Signed Numbers

There are two situations that arise in adding signed numbers:

Case 1: If the signs are the *same, add* their absolute values and *keep* the sign. Examples:

$$^+4 + {}^+11 = +(4 + 11) = +15$$
$$^-21 + {}^-13 = -(21 + 13) = -44$$

Case 2: If the signs are *different, subtract* their absolute values (smaller from the larger) and *take the sign of the number with the larger absolute value.* Examples:

$$^+5 + {}^-8 = -(8 - 5) = -3$$
$$^+32 + {}^-18 = +(32 - 18) = +14$$

142

It is helpful to think of signed numbers as representing temperatures rising or falling to see if your answer makes sense.

Key Fact: *The sum of opposites is 0.* Example:

$$^-15 + {}^+15 = \pm(15 - 15) = 0$$

Subtracting Signed Numbers

Subtracting a signed number is the same as adding its opposite. Think of temperature, for example. If the temperature is $+80°F$ and it falls by $20°F$, the problem is $^+80 - {}^+20$. However, "falling by $20°F$" is the same as saying the temperature has become $20°F$ less, which is the problem $^+80 + {}^-20$.

Therefore, after making the change from subtraction to addition, you can follow the rules for addition. Consider the following examples:

$$^+15 - {}^+4 = {}^+15 + {}^-4 = +11$$
$$^-8 - {}^+5 = {}^-8 + {}^-5 = -13$$
$$^-12 - {}^-7 = {}^-12 + {}^+7 = -5$$

Key Fact: *Any signed number subtracted from itself is 0.* Example:

$$^-21 - {}^-21 = {}^-21 + {}^+21 = 0$$

Multiplying and Dividing Signed Numbers

As described below, there are two cases for finding the product and quotient of signed numbers.

Case 1: If the signs are the *same*, the product or quotient is *positive*. Examples:

$$-8 \times -2 = +16$$
$$-21 \div -3 = +7$$

Case 2: If the signs are *different*, the product or quotient is *negative*. Examples:

$$+3 \times -2 = -6$$
$$-18 \div +6 = -3$$
$$+5 \times -1 = -5$$

Key Fact: *The reciprocal of a signed number must have the same sign.* This is true because the product of reciprocals must be +1. Examples:

$$+3 \times +\frac{1}{3} = +1$$
$$-\frac{2}{3} \times -\frac{3}{2} = +1$$

Key Fact: *Any signed number is the product of −1 and its opposite.* Examples:

$$-7 \times -1 = +7$$
$$+6 \times -1 = -6$$

A Brief Review of Exponents

The information below will refresh your memory about some of the most important facts about exponents.

Working with Exponents

A number such as 3^4 is referred to as *the fourth power of 3*, where 3 is the *base* and 4 is the *exponent*. Exponents indicate repeated multiplication, and, therefore, 3^4 is *the product of four 3's* or $3 \times 3 \times 3 \times 3 = 81$. Other examples are the following:

$$4^2 = 4 \times 4 = 16 \qquad 2^4 = 2 \times 2 \times 2 \times 2 = 16$$
$$5^3 = 5 \times 5 \times 5 = 125 \qquad 3^5 = 3 \times 3 \times 3 \times 3 \times 3 = 243$$

In many problems you may find yourself working with powers of variables. The same definition applies. Examples are $x^3 = xxx$ and $y^5 = yyyyy$.

In applications, we often see exponents when finding area or volume.

Example. The area of a square whose side measures 7 in is $7^2 = 7 \text{ in} \times 7 \text{ in} = 49 \text{ in}^2$.

Example. The volume of a cube whose edge measures 7 in is $7^3 = 7 \text{ in} \times 7 \text{ in} \times 7 \text{ in} = 343 \text{ in}^3$.

Because of these applications, special names are given to numbers with exponents of 2 and 3. The integer x^2 is referred to as *x squared* and x^3 is referred to as *x cubed*. There are no special names for other powers.

Exponents can be used with any kind of number, including fractions, decimals, and signed numbers. Examples are $(^2\!/_3)^2 = {}^2\!/_3 \times {}^2\!/_3 = {}^4\!/_9$ and $(1.5)^3 = 1.5 \times 1.5 \times 1.5 = 3.375$; also

$$\left(\frac{x}{y}\right)^4 = \left(\frac{x}{y}\right)\left(\frac{x}{y}\right)\left(\frac{x}{y}\right)\left(\frac{x}{y}\right) = \frac{x^4}{y^4}$$

Additional examples are $(-2)^3 = (-2)(-2)(-2) = -8$ and $(-1)^4 = (-1)(-1)(-1)(-1) = +1$.

Key Fact: *An odd power of a negative number is negative. An even power of a negative number is positive.*

Special Powers

$x^1 = x$	A number raised to the *first power* is itself. Examples: $4^1 = 4$, $(-5)^1 = -5$, and $(^1\!/_2)^1 = {}^1\!/_2$.
$x^0 = 1$	A number raised to the *zero power* is 1. Examples: $4^0 = 1$, $(-5)^0 = 1$, and $(^1\!/_2)^0 = 1$.
$x^{-1} = 1/x$	A number raised to the -1 *power* is its reciprocal. Examples: $4^{-1} = {}^1\!/_4$, $(-5)^{-1} = -{}^1\!/_5$, and $(^1\!/_2)^{-1} = 2$.
$x^{-2} = 1/x^2$, $x^{-3} = 1/x^3$, $x^{-4} = 1/x^4$, etc.	A number raised to any negative power is the reciprocal of the number raised to its positive counterpart. Examples: $4^{-2} = (^1\!/_4)^2 = {}^1\!/_{16}$, $(-5)^{-3} = (^1\!/_5)^3 = {}^1\!/_{125}$, and $(^1\!/_2)^{-4} = 2^4 = 16$.

Multiplying and Dividing Powers of the Same Base

When multiplying and dividing numbers with exponents, we can apply the *laws of exponents.*

I. $x^a x^b = x^{a+b}$	The *product* of powers of the same base is that base raised to the *sum of the exponents.* Examples: $5^3\,5^2 = 5^{3+2} = 5^5$, $(-3)^4\,(-3)^5 = (-3)^{4+5} = (-3)^9$, and $y^4 y = y^{4+1} = y^5$.
II. $x^a/x^b = x^{a-b}$	The *quotient* of powers of the same base is that base raised to the *difference of the exponents.* Examples: $5^3/5^2 = 5^{3-2} = 5^1$, $(-3)^4/(-3)^5 = (-3)^{4-5} = (-3)^{-1}$, and $y^4/y = y^{4-1} = y^3$.

146

III. $(x^a)^b = x^{ab}$ *A power of a base raised to another power* is that base raised to the *product of the exponents*. Examples: $(5^3)^2 = 5^{3\times2} = 5^6$, $[(-3)^4]^5 = (-3)^{4\times5} = (-3)^{20}$, and $(y^4)^1 = y^{4\times1} = y^4$.

Key Fact: These rules do <u>not</u> apply to multiplying or dividing <u>different</u> *powers of different bases*.

     Example. $2^4 3^5 = 16 \times 243 = 3{,}888$.

     Example. $x^4 y^5$ is not combined with any other form.

Key Fact: We do <u>not</u> combine exponents when <u>adding or subtracting</u> *powers of the same or different bases*.

     Example. $2^4 + 3^5 = 16 + 243 = 259$.

     Example. $x^4 + y^5$ is not combined with any other form.

A Brief Review of Radicals

The information below will refresh your memory about some of the most important facts about radical expressions and roots.

A *radical* is an expression that represents a root of a number. The most common root of a number is its *square root*. The relationship is (square root of a number)^2 = number. The symbol $\sqrt{}$ is used to represent this root.

Example. $\sqrt{2}$ means the *square root of 2*, which is approximately 1.41421356237 according to the calculator, and $(\sqrt{2})^2 = 2$.

Example. $\sqrt{10}$ means the *square root of 10*, which is approximately 3.16227766017 according to the calculator, and $(\sqrt{10})^2 = 10$.

Radical x is another way of referring to the square root of x or $\sqrt{x}$.

Key Fact: $(\sqrt{x})^2 = \sqrt{x}\sqrt{x} = x$ *for all positive numbers and zero.*

Key Fact: $\sqrt{0} = 0$, *and* $\sqrt{1} = 1$.

Key Fact: *A negative number does <u>not</u> have a square root!* (At least not in the set of real numbers we usually work with.) You cannot multiply a signed number by itself and get a negative number.

Example. $\sqrt{-4} \neq -2$ or $+2$ since $(-2)(-2) = +4$ and $(+2)(+2) = +4$.

148

Key Fact: *The square root of a proper fraction is <u>larger</u> than the fraction. Examples:*

$$\sqrt{\frac{1}{4}} = \frac{1}{2} \quad \text{since} \quad \left(\frac{1}{2}\right)\left(\frac{1}{2}\right) = \frac{1}{4}$$

$$\sqrt{\frac{9}{16}} = \frac{3}{4} \quad \text{since} \quad \left(\frac{3}{4}\right)\left(\frac{3}{4}\right) = \frac{9}{16}$$

Key Fact: $\sqrt{x^2} = x$ *for any positive number x. Example:*

$$\sqrt{5^2} = \sqrt{25} = 5$$

Multiplying and Simplifying Radicals

The product of the roots of two numbers is the same as the root of the product of the numbers and vice versa. For instance, $\sqrt{x}\sqrt{y}$ is the same as $\sqrt{x\,y}$.

Example. $\sqrt{4}\sqrt{9} = \sqrt{4 \times 9} = \sqrt{36} = 6$. This, of course, is the same as $\sqrt{4}\sqrt{9} = 2 \times 3 = 6$.

Example. $\sqrt{5}\sqrt{20} = \sqrt{5 \times 20} = \sqrt{100} = 10$.

Example. $\sqrt{72} = \sqrt{36 \times 2} = \sqrt{36}\sqrt{2} = 6\sqrt{2}$. This is the way you *simplify a radical* so that the expression has the *simplest root.*

Example. $\sqrt{147} = \sqrt{49 \times 3} = \sqrt{49}\sqrt{3} = 7\sqrt{3}$.

Example. $\sqrt{x^3} = \sqrt{x^2}\sqrt{x} = x\sqrt{x}$.

Dividing Radicals

The quotient of the roots of two numbers is the same as the root of the quotient of the numbers and vice versa. For instance, $\sqrt{x}/\sqrt{y}$ is the same as $\sqrt{x/y}$. Consider the following examples:

$$\frac{\sqrt{4}}{\sqrt{9}} = \sqrt{\frac{4}{9}} = \frac{2}{3} \qquad \frac{\sqrt{5}}{\sqrt{20}} = \sqrt{\frac{5}{20}} = \sqrt{\frac{1}{100}} = \frac{\sqrt{1}}{\sqrt{100}} = \frac{1}{10}$$

Roots Other Than the Square Root

There are roots other than the square root. For example, there are *cube roots, fourth roots, fifth roots*, and so on. They are represented by the radical sign with a smaller number indicating which root it is.

Example. $\sqrt[3]{8}$ is *the cube root of 8*, and it represents the number which, when raised to the third power, is 8. The number is 2 since $2^3 = 8$.

Example. $\sqrt[4]{81}$ is *the fourth root of 81*, and it represents the number which, when raised to the fourth power, is 81. The number is 3 since $3^4 = 81$

Key Fact: *The rules for multiplying, simplifying, and dividing square roots hold for all other roots, as long as you are working with the <u>same</u> root!* Example:

$$\sqrt[3]{27 \times 8} = \sqrt[3]{27}\sqrt[3]{8} = 3 \times 2 = 6$$

Key Fact: We do <u>not</u> combine roots when <u>adding or subtracting</u>. Examples:

$\sqrt{9} + \sqrt{16} = 3 + 4 = 7$

$\sqrt{9} + \sqrt{16} \neq \sqrt{9+16} = \sqrt{25} = 5!!!$

$\sqrt{3} + \sqrt{3} = 2\sqrt{3} \approx 2 \times 1.73 = 3.46$

$\sqrt{3} + \sqrt{3} \neq \sqrt{6} \approx 2.45$

Example. $\sqrt{x} + \sqrt{y}$ is not combined with any other form.

A Brief Review of Polynomials

The information below will refresh your memory about some of the most important facts about polynomials.

In algebra, an expression containing whole-number powers of variables is called a *polynomial*. Each part of the polynomial separated by + or − is called a *term*.

$x^2 + 2x + 7$ The three terms are x^2, $+2x$, and -7. A three-term polynomial is called a *trinomial*.

$-y^3 - 5y^2 + 2y - 1$ The four terms are $-y^3$, $-5y^2$, $+2y$, and -1.

$x^2y^3 + 7xy - 3y^2$ The three terms are x^2y^3, $+7xy$, and $-3y^2$.

$x^2 - 9$ The two terms are x^2 and -9. A two-term polynomial is called a *binomial*.

$8a^3b$ The one term is $8a^3b$. A one-term polynomial is called a *monomial*.

The most common tasks performed on polynomials are *adding, subtracting, multiplying*, and *factoring*.

Key Fact: *To add or subtract polynomials, combine the terms that are exactly identical.* Examples:

$$(4x^2y - 2xy^2 + 5) + (6x^2y - 3) = 10x^2y - 2xy^2 + 2$$
$$(9x^2 + 5x + 4) - (4x^2 + 2x - 7) = 5x^2 - 3x + 11$$

Key Fact: To multiply a monomial by a polynomial, use the distributive property and the laws of exponents. Examples:

$$2x^3(x^2 + x - 3) = 2x^3x^2 + 2x^3x - 2x^3(3)$$
$$= 2x^5 + 2x^4 - 6x^3$$
$$-3y(10 - 3y + y^2) = (-3y)(10) + (-3y)(-3y) + (-3y)y^2$$
$$= -30y + 9y^2 - 3y^3$$

Multiplying Two Binomials

In word problems, the most common, and probably the most important, multiplication of polynomials you are asked to perform is to multiply two binomials. To do this, there are many methods. The one you are probably most familiar with is *FOIL*—that is, identify the pair of *first terms*, the pair of *outside terms*, the pair of *inside terms*, and the pair of *last terms*.

Key Fact: The product of the two binomials is the sum of the products of the terms in each pair. Example: $(x + 3)(2x - 7)$.

The *first terms* in each binomial are x and $2x$. Their product is $2x^2$.

The *outside terms* from each binomial are x and -7. Their product is $-7x$.

The *inside terms* from each binomial are $+3$ and $2x$. Their product is $+6x$.

The *last terms* in each binomial are $+3$ and -7. Their product is -21.

Therefore, $(x + 3)(2x - 7) = 2x^2 - 7x + 6x - 21$, and the middle terms combine to give us $2x^2 - x - 21$.

A second example: $(x - 5)^2 = (x - 5)(x - 5)$.

The *first terms* in each binomial are x and x. Their product is x^2.

The *outside terms* from each binomial are x and -5. Their product is $-5x$.

The *inside terms* from each binomial are -5 and x. Their product is $-5x$.

152

The *last terms* in each binomial are -5 and -5. Their product is $+25$.

Therefore, $(x - 5)^2 = x^2 - 5x - 5x + 25$, and the middle terms combine to give us $x^2 - 10x + 25$.

A third example: $(x + 8)(x - 8)$.

The *first terms* in each binomial are x and x. Their product is x^2.

The *outside terms* from each binomial are x and -8. Their product is $-8x$.

The *inside terms* from each binomial are $+8$ and x. Their product is $+8x$.

The *last terms* in each binomial are $+8$ and -8. Their product is -64.

Therefore, $(x + 8)(x - 8) = x^2 - 8x + 8x - 64$, and *the middle terms combine to 0*, giving us $x^2 - 64$.

The polynomials in each of the three examples above are referred to as *quadratic expressions* because in each, the highest power of the variable is 2.

Factoring a Quadratic

When given a quadratic trinomial, you often have to find the two binomials whose product is the quadratic. To accomplish this, you have to reverse the process used to multiply the binomials. This requires finding the first and last terms of the two binomials so that the four products combine to give you the quadratic.

Example. Factor $x^2 + 8x + 15$ into the product of two binomials. The most logical first terms are x and x: $(x \quad)(x \quad)$. Since the last term is *positive*, the signs should be the *same*. The possibilities are $(x + \quad)(x + \quad)$ or $(x - \quad)(x - \quad)$. The *product of the last terms has to be +15*, and *the sum of the products of the inside and outside terms has to be +8x*. Therefore, we need the cofactors of 15 with the same sign whose sum is $+8$. The only possibilities are $+5$ and $+3$. The factorization must be $(x + 5)(x + 3)$, and, if you use FOIL, you can verify that it is correct.

Example. Factor $x^2 - 12x + 32$ into the product of two binomials. The most logical first terms are x and x $(x \quad)(x \quad)$. Since the last term is *positive*, the signs should be the *same*. The possibilities are $(x + \quad)(x + \quad)$ or $(x - \quad)(x - \quad)$. The *product of the last terms has to be +32*, and *the sum of the products of the inside and outside terms has to be $-12x$*. Therefore, we need the cofactors of 32 with the same sign whose sum is -12. The only possibilities are -8 and -4. The factorization must be $(x - 8)(x - 4)$, and, if you use FOIL, you can verify that it is correct.

Example. Factor $x^2 + 8x - 20$ into the product of two binomials. The most logical first terms are x and x $(x \quad)(x \quad)$. Since the last term is *negative*, the signs should be the *different* $(x + \quad)(x - \quad)$. *The product of the last terms has to be -20*, and *the sum of the products of the inside and outside terms has to be $+8x$*. Therefore, we need the cofactors of 20 with the different signs whose sum is $+8$. The only possibilities are $+10$ and -2. The factorization must be $(x + 10)(x - 2)$, and, if you use FOIL, you can verify that it is correct.

Example. Factor $x^2 - 3x - 28$ into the product of two binomials. The most logical first terms are x and x: $(x \quad)(x \quad)$. Since the last term is *negative*, the signs should be the *different* $(x + \quad)(x - \quad)$. *The product of the last terms has to be -28*, and *the sum of the products of the inside and outside terms has to be $-3x$*. Therefore, we need the cofactors of 28 with the different signs whose sum is -3. The only possibilities are $+4$ and -7. The factorization must be $(x + 4)(x - 7)$, and, if you use FOIL, you can verify that it is correct.

Example. Factor $x^2 - 36$ into the product of two binomials. It appears that we are missing a middle term. No problem! We can always add 0 or, in this case, $0x$ to give us the complete trinomial:

$x^2 + 0x - 36$. The most logical first terms are x and x $(x\ )(x\ )$. Since the last term is *negative*, the signs should be the *different* $(x +\ )(x -\ )$. *The product of the last terms has to be* -36, and *the sum of the products of the inside and outside terms has to be* $0x$. Therefore, we need the cofactors of 36 with the different signs whose sum is 0. The only possibilities are $+6$ and -6. The factorization must be $(x + 6)(x - 6)$, and, if you use FOIL, you can verify that it is correct.

Key Fact: *The factorization of the difference of two perfect squares,* $x^2 - a^2$, *is* $(x + a)(x - a)$. Examples:

$$x^2 - 100 = (x + 10)(x - 10)$$
$$y^2 - 144 = (y + 12)(y - 12)$$

A Brief Review of Solving Equations

The information below will refresh your memory about some of the most important facts about solving equations.

The two most common types of equations that you need to solve are *linear equations* and *quadratic equations*. For each there are several steps to follow that will ensure that you get the correct answer every time.

Linear Equations

An equation is *linear* if the highest power of the variable is 1 such as $4x - 5 = 11$. In fact, you won't even see the power since we rarely write exponents of 1. The goal is to undo all the complications in the equations and work your way to having the variable appear on only one side, with a numerical term on the other. The following example illustrates the necessary steps involved:

Example. Solve the equation $4(2x + 5) + 6x = 8x - 4 - 2x$ for the value of x.

Step 1

Use the distributive property to "remove" any parentheses in the equation.

$$8x + 20 + 6x = 8x - 4 - 2x$$
$$14x + 20 = 6x - 4$$

Step 2

Combine like terms on each side of the equation separately.

At this point the variable should appear at most once on each side, and a number should appear at most once on each side.

Step 3

Remove the variable with the smaller coefficient by adding its signed opposite to both sides of the equation, and combine the variable terms.

$$14x + (-6x) + 20 = 6x + (-6x) - 4$$
$$8x + 20 = -4$$

Step 4

Remove the numerical term from the side of the equation with the variable by adding its signed opposite to both sides of the equation, and combine the numerical terms.

$$8x + 20 + (-20) = -4 + (-20)$$
$$8x = -24$$

Step 5

Divide both sides of the equation by the coefficient of the variable and simplify.

$$8x \div 8 = -24 \div 8$$
$$x = -3$$

Step 6

Check your answer in the original equation by using it in place of the variable.

$$4(2x + 5) + 6x = 8x - 4 - 2x$$
$$4(2 \times (-3) + 5) + 6 \times (-3) = 8 \times (-3) - 4 - 2 \times (-3)$$
$$4(-6 + 5) + -18 \qquad -24 - 4 + 6$$
$$4 \times (-1) + -18 \qquad -22$$
$$-4 + -18$$
$$-22 \qquad\qquad\qquad \checkmark$$

Example.   Solve the equation $4(y - 7) - 2(8 - y) = 6(y + 3) - 3y + 1$.

Step 1

Use the distributive property to "remove" any parentheses in the equation.

$$4y - 28 - 16 + 2y = 6y + 18 - 3y + 1$$

Step 2

Combine like terms on each side of the equation separately.

$$6y - 44 = 3y + 19$$

Step 3

Remove the variable with the smaller coefficient by adding its signed opposite to both sides of the equation, and combine the variable terms.

$$6y + (-3y) - 44 = 3y + (-3y) + 19$$
$$3y - 44 = 19$$

Step 4

Remove the numerical term from the side of the equation with the variable by adding its signed opposite to both sides of the equation, and combine the numerical terms.

$$3y - 44 + 44 = 19 + 44$$
$$3y = 63$$

Step 5

Divide both sides of the equation by the coefficient of the variable and simplify.

$$3y \div 3 = 63 \div 3$$
$$y = 21$$

Step 6

Check your answer in the original equation by using it in place of the variable.

(Try the check yourself. Each side has a value of 82 when $y = 21$.)

158

Key Fact: You can add, subtract, multiply, or divide only if you perform the same operation on *each side* of the equation.

Simultaneous Linear Equations with Two Variables

A pair of equations is said to be a *simultaneous* pair if you are told to look for the values of each variable that satisfy *both* of the equations. Only one pair of numbers will work. For example, the pair of numbers $x = 7$ and $y = 2$ is the only pair that will satisfy the equations $x + y = 9$ and $x - y = 5$ at the same time. We usually write the solution as an *ordered pair* (x, y). In this case the solution is $(7, 2)$. Simultaneous equations arise frequently in word problems that involve two unknowns. There are many methods for solving simultaneous pairs of equations. The two most common methods are by substitution and by combining.

Substitution is usually used when one equation lets you easily write one variable in terms of the other.

Example. Solve the following pair of simultaneous equations for x and y:

$$2x + y = 6 \qquad y - 3x = -14$$

Step 1

Solve one of the equations for one of the variables.

The second equation is easily solved for y by adding $3x$ to both sides, giving $y = 3x - 14$.

Step 2

Use the expression for the solved variable in its place in the other equation.

Substituting $3x - 14$ for y in the first equation gives $2x + 3x - 14 = 6$.

Step 3

Solve the "new" equation for its variable.

Combining and solving for x, we have

$$5x - 14 = 6$$
$$5x = 20$$
$$x = 4$$

Use this variable to find the other variable from the equation you used in step 1.

$$y - 3x = -14$$
$$y - 3(4) = -14$$
$$y - 12 = -14$$
$$y = -2$$

Step 5
Write your solution.

The solution is the pair $x = 4$ and $y = -2$ or, as an ordered pair (x, y), $(4, -2)$.

Step 6
Check that your solution satisfies both equations.

In the first equation

$$2(4) + (-2) = 6$$
$$8 + (-2) \quad \checkmark$$
$$6$$

In the second equation

$$-2 - 3(4) = -14$$
$$-2 - 12$$
$$-14 \qquad \checkmark$$

Example. Solve the following pair of simultaneous equations for a and b:

$$3a - b = 20 \qquad 2a + b = 15$$

Step 1
Solve one of the equations for one of the variables.

The second equation is easily solved for b by adding $-2a$ to both sides, giving $b = -2a + 15$.

160

Step 2

Use the expression for the solved variable in its place in the other equation.

Substituting $-2a + 15$ for b in the first equation gives $3a - (-2a + 15) = 20$.

Step 3

Solve the "new" equation for its variable.

Combining and solving for x, we have

$$5a - 15 = 20$$
$$5a = 35$$
$$a = 7$$

Step 4

Use this variable to find the other variable from the equation you used in step 1.

$$2a + b = 15$$
$$14 + b = 15$$
$$b = 1$$

Step 5

Write your solution.

The solution is the pair $a = 7$ and $b = 1$ or, as an ordered pair (a, b), $(7, 1)$.

Step 6

Check that your solution satisfies both equations.

In the first equation

$$3(7) - 1 = 20$$
$$21 - 1$$
$$20 \quad \checkmark$$

In the second equation

$$2(7) + 1 = 15$$
$$14 + 1$$
$$15 \quad \checkmark$$

Solving by *combining* is sometimes easier as it immediately eliminates one of the variables. Using the last example, we can add the given equations to eliminate b.

$$3a - b = 20$$
$$\underline{+2a + b = 15}$$
$$5a = 35$$

and we immediately see that $a = 7$. We can continue by using this value in one of the equations to find the value of the other variable as in step 4 above.

If the equations can't be added to eliminate one of the variables, we can "fix" the situation by doing a bit of multiplying first.

Example.   Solve the following pair of simultaneous equations for x and y:

$$3x + 2y = 36 \qquad 4x - 5y = 2$$

Step 1
Select one of the variables to eliminate and multiply each equation by the coefficient of this variable from the other equation.

If we choose to eliminate x, we will multiply the first equation by 4 and the second by 3, giving us the pair $12x + 8y = 144$ and $12x - 15y = 6$.

Step 2
Multiply both sides of one of these "new" equations by -1.

Multiplying both sides of the second equation by -1 gives us the pair $12x + 8y = 144$ and $-12x + 15y = -6$.

Step 3
Add the equations to eliminate the variable, and solve the resulting equation for the other variable.

Adding the two equations gives us $23y = 138$ and $y = 138 \div 23 = 6$.

Step 4
Use this value in either of the two original equations to find the value of the other.

Using $y = 6$ in the first equation, we have $3x + 12 = 36$, $3x = 24$, and $x = 8$.

162

Step 5

Write your solution.

The solution is the pair $x = 8$ and $y = 6$ or, as an ordered pair (x, y), $(8, 6)$.

Step 6

Check that your solution satisfies both equations.

(Try the check yourself as practice.)

Quadratic Equations

In a *quadratic equation*, the highest power of the variable is 2. For example, the following are quadratic equations:

$x^2 = 25$, $y^2 - 5y = 0$, and $n^2 - 6n + 8 = 0$.

Key Fact: Every quadratic equation can be solved for *two values* of the variable.

Example. $x^2 = 25$ has the obvious solutions $x = 5$ and $x = -5$.

Example. $y^2 - 5y = 0$ has the solutions $y = 0$ and $y = 5$.

Example. $n^2 - 6n + 8 = 0$ has the solutions $n = 2$ and $n = 4$. We sometimes find that these two solutions are equal.

Example. $x^2 - 8x + 16 = 0$ has the solutions $x = 4$ and $x = 4$.

The most common method of solving a quadratic equation is *by factoring* using the steps described below.

Example. Solve for all values of x that satisfy $x^2 = 25$.

Step 1

Rewrite the equation so that one side is 0.

This is accomplished by adding -25 to both sides: $x^2 + (-25) = 25 + (-25)$ or $x^2 - 25 = 0$.

Step 2

Factor the quadratic expression.

This particular quadratic is a difference of two perfect squares, which factors as $(x + 5)(x - 5) = 0$.

The next step uses the fact that if the product of two expressions is 0, then one of the factors must be 0. (How else could you multiply two numbers to get 0?)

Step 3

Set each factor equal to 0, and solve for the variable.

We have the two equations, $x + 5 = 0$ and $x - 5 = 0$, which have the solutions $x = -5$ and $x = +5$.

Step 4

Check each solution in the original equation.

Clearly, $x^2 = 25$ when x is $+5$ or -5. ✓

Example. Solve for all values of y that satisfy $y^2 - 5y = 0$.

Step I

Rewrite the equation so that one side is 0.

This is already the case: $y^2 - 5y = 0$.

Step 2

Factor the quadratic expression.

Both terms of the particular quadratic have the common factor y, so it factors as $y(y - 5) = 0$.

Step 3

Set each factor equal to 0, and solve for the variable.

We have the two equations, $y = 0$ and $y - 5 = 0$, which have the solutions $y = 0$ and $y = +5$.

Step 4

Check each solution in the original equation.

Using $y = 0$, we have

$$0^2 - 0(5) = 0$$
$$0 - 0$$
$$0 \quad ✓$$

Using $y = 5$, we have

$$5^2 - 5(5) = 0$$
$$25 - 25$$
$$0 \quad ✓$$

Example. Solve for all values of n that satisfy $n^2 + 8 = 6n$.

164

Step 1

Rewrite the equation so that one side is 0.

This is done by adding $-6n$ to both sides, giving us $n^2 - 6n + 8 = 0$.

Step 2

Factor the quadratic expression.

This particular quadratic factors as $(n - 2)(n - 4) = 0$.

Step 3

Set each factor equal to 0, and solve for the variable.

We have the two equations, $n - 2 = 0$ and $n - 4 = 0$, which have the solutions $n = +2$ and $n = +4$.

Step 4

Check each solution in the original equation.

Using $n = 2$, we have

$$2^2 + 8 = 6(2)$$
$$4 + 8 \quad\quad 12$$
$$12 \quad\quad \checkmark$$

Using $n = 4$, we have

$$4^2 + 8 = 6(4)$$
$$16 + 8 \quad\quad 24$$
$$24 \quad\quad \checkmark$$

Example.   Solve for all values of x that satisfy $x^2 = 8x - 16$.

Step 1

Rewrite the equation so that one side is 0.

This is done by adding $-8x$ and $+16$ to both sides, giving us $x^2 - 8x + 16 = 0$.

Step 2

Factor the quadratic expression.

This particular quadratic factors as $(x - 4)(x - 4) = 0$.

Step 3

Set each factor equal to 0, and solve for the variable.

We have the two equations, $x - 4 = 0$ and $x - 4 = 0$, which have the equal solutions $x = 4$.

Step 4
Check each solution in the original equation.
Using $x = 4$, we have

$$4^2 = 8(4) - 16$$
$$16 \qquad 32 - 16$$
$$16 \checkmark$$

Key Fact: There are quadratic expressions that do not factor. To solve these, you need to apply the following special *quadratic formula*. The quadratic expression $ax^2 + bx + c = 0$ has the two solutions:

$$x = \frac{-b \pm \sqrt{b^2 - 4ac}}{2a}$$

(where $a, b,$ and c represent the numbers that appear in the expression in those specific places; i.e., they are the *coefficients* of the terms).

This is beyond the scope of this book, but the following example will show how it can be used. The equation $x^2 - 9x - 22 = 0$ has the coefficients $a = 1, b = -9,$ and $c = -22$. These values are substituted into the formula, and we have

$$x = \frac{-(-9) \pm \sqrt{(-9)^2 - 4(1)(-22)}}{2(1)}$$

$$= \frac{9 \pm \sqrt{81 + 88}}{2} = \frac{9 \pm \sqrt{169}}{2} = \frac{9 \pm 13}{2}$$

$$= \frac{9 + 13}{2} = \frac{22}{2} = 11 \qquad x = \frac{9 - 13}{2} = \frac{-4}{2} = -2$$

You should check these values to verify that they are solutions. Also, this quadratic, $x^2 - 9x - 22 = 0$, can be solved by factoring, and you should do so as practice.

Index

The index below includes the most relevant items found within the text, examples, problems and solutions. If the item is found within a problem or a solution, the chapter number and problem number appear in parentheses after the page reference. For example, if we refer to "Age" below, we see that "age" is a major component in Problem 7 of Chapter 2 on page 36 as well as in two other problems. Subitems are included to be more specific about the item where appropriate.

Absolute Value, 142
Age, 36 (2-[Prob.7]), 42 (2-[Prob.7]), 121 (6-[Prob.14])
Angles, 112 (5-[Prob.10]), 115 (5-[Prob.10]), 119 (6-[Prob.5])
  Adjacent, 133
  alternate interior, 45 (2-[Prob.10])
  base, 10
  bisector, 133
  complementary, 9 (1-[Ex.10]), 133, 134
  exterior, 30 (2-[Ex.9]), 134
  supplementary, 30 (2-[Ex.9]), 133, 134
  vertex, 10
  vertical, 133, 134
Area, 62 (3-[Prob.8]), 68 (3-[Prob.8]), 109 (5-[Ex.10]), 121 (6-[Prob.15]), 121 (6-[Prob.16])
  formulas, 130–131
  measures, 128
Averages, 4 (1-[Ex.4]), 10 (1-[Ex.11]), 32 (2-[Ex.11]), 37 (2-[Prob.12]), 46 (2-[Prob.12]), 107 (5-[Ex.6]), 108 (5-[Ex.7]), 111 (5-[Prob.5]), 114 (5-[Prob.5]), 118 (6-[Prob.2]), 124 (6-[Prob.28])

Circles, 112 (5-[Prob.11]), 112 (5-[Prob.11]), 121 (6-[Prob.16]), 125 (6-[Prob.34]), 126 (6-[Prob.36]), 136
  area, 13 (1-[Prob.13]), 19 (1-[Prob.13]), 30 (2-[Ex.10]), 62 (3-[Prob.11]), 70 (3-[Prob.11]), 131

circumference, 13 (1-[Prob.12]), 13 (1-[Prob.13]), 19 (1-[Prob.13]), 19 (1-[Prob.12]), 130
diameter, 19 (1-[Prob.13]), 30 (2-[Ex.10]), 116 (5-[Prob.11])
Coins, 36 (2-[Prob.8]), 43 (2-[Prob.8])
Cone, 63 (3-[Prob.12]), 71 (3-[Prob.12])
  volume, 132
Counting, 92 (4-[Prob.10])), 92 (4-[Prob.6])), 98 (4-[Prob.6])), 101 (4-[Prob.10])), 112, 112 (5-[Prob.12])), 115 (5-[Prob.9])), 116 (5-[Prob.12])), 123 (6-[Prob.26])), 124 (6-[Prob.29])), 124 (6-[Prob.30])), 125 (6-[Prob.33])), 126 (6-[Prob.37])
  arrangements, 74 (4-[Ex.1]), 76 (4-[Ex.2]), 91 (4-[Prob.1]), 92 (4-[Prob.1])
  counting principle, 75 (4-[Ex.1]), 92 (4-[Prob.1])
  sets, 82 (4-[Ex.5])
Cylinder, 131

Data collection:
  pie graph, 123 (6-[Prob.24])
  tables, 32 (2-[Ex.11])
Difference of squares, 5 (1-[Ex.5]), 12 (1-[Prob.5]), 15 (1-[Prob.5]), 111 (5-[Prob.4]), 113 (5-[Prob.4]), 155
Digit, 2 (1-[Ex.1]), 3 (1-[Ex.3]), 112 (5-[Prob.12]), 116 (5-[Prob.12]), 124 (6-[Prob.29]), 125 (6-[Prob.31])

Distance, 12 (1-[Prob.6]), 12
  (1-[Prob.7]), 16 (1-[Prob.6]), 24
  (2-[Ex.4]), 35 (2-[Prob.4]), 39
  (2-[Prob.4]), 110 (5-[Ex.11])
  formula, 130
  measures, 128
Distributive law, 37

Equations:
  combining, 14 (1-[Prob.2])
  linear, 156–159
  quadratics, 163–166
  simultaneous, 113 (5-[Prob.3]), 119
    (6-[Prob.7]), 125 (6-[Prob.35]),
    159–163
Exponents, 28 (2-[Ex.7])
  review, 145–147

Factoring, 20 (1-[Prob.15]), 91
  (4-[Prob.3]), 92 (4-[Prob.10]), 94
  (4-[Prob.3]), 101 (4-[Prob.10]), 105
  (5-[Ex.2]), 126 (6-[Prob.38])
  quadratics, 153–155
Formulas, table of 130–132
Fractions, 6, 7 (1-[Ex.7]), 8 (1-[Ex.9]),
  12 (1-[Prob.1]), 35 (2-[Prob.2]), 38
  (2-[Prob.2]), 97 (4-[Prob.5]), 109
  (5-[Ex.9]), 109 (5-[Ex.8]), 111
  (5-[Prob.7]), 114 (5-[Prob.7]), 119
  (6-[Prob.4]), 126 (6-[Prob.40])
  adding, 140
  algebraic, 8, 22 (2-[Ex.2]), 28
    (2-[Ex.8]), 38 (2-[Prob.2]), 41
    (2-[Prob.6]), 43 (2-[Prob.7]), 49
    (3-[Ex.3])
  dividing, 139
  equivalent, 138
  improper, 137
  mixed numbers, 137
  multiplying, 138
  reducing, 138
  review, 137–140
  subtracting, 140

Geometry, definitions and concepts
  133–136
Gridding answers, 76 (4-[Ex.2]), 78
  (4-[Ex.3]), 79 (4-[Ex.4]), 85
  (4-[Ex.8]), 86 (4-[Ex.9]), 87
  (4-[Ex.11]), 89 (4-[Ex.12]), 90
  (4-[Ex.13]), 93 (4-[Prob.2]), 95
  (4-[Prob.4]), 97 (4-[Prob.5]), 99
  (4-[Prob.6]), 101 (4-[Prob.10])

Inscribed figures, 115, 136
Integers, 3 (1-[Ex.2]), 13 (1-[Prob.14]),
  19 (1-[Prob.14]), 63 (3-[Prob.14]),

72 (3-[Prob.14]), 85 (4-[Ex.8]), 86
  (4-[Ex.9]), 86 (4-[Ex.10]), 87
  (4-[Ex.11]), 92 (4-[Prob.8]), 100
  (4-[Prob.8]), 105 (5-[Ex.1]), 111
  (5-[Prob.1]), 112 (5-[Prob.1]), 119
  (6-[Prob.3]), 120 (6-[Prob.8]), 123
  (6-[Prob.25]), 125 (6-[Prob.32]), 126
  (6-[Prob.39])
  consecutive, 111 (5-[Prob.5]), 114
    (5-[Prob.5])
  definition, 3

Line segments, 36 (2-[Prob.9]), 44
  (2-[Prob.9])
Lists:
  arrangements, 75, 84 (4-[Ex.7]), 93
    (4-[Prob.2]), 115 (5-[Prob.9]), 116
    (5-[Prob.12])
  flowchart, 83 (4-[Ex.6])
  multiples, 88 (4-[Ex.12]), 102
    (4-[Prob.11])
  outcomes, 78 (4-[Ex.3]), 101
    (4-[Prob.9]), 108 (5-[Ex.7])
  using tables, 77 (4-[Ex.3]), 80
    (4-[Ex.4]), 94 (4-[Prob.3]), 95
    (4-[Prob.4]), 99 (4-[Prob.7]), 101
    (4-[Prob.10])
Logical reasoning, 82 (4-[Ex.6]), 84
  (4-[Ex.7]), 91 (4-[Prob.4]), 92
  (4-[Prob.7]), 94 (4-[Prob.4]), 99
  (4-[Prob.7]), 112 (5-[Prob.12]), 116
  (5-[Prob.12]), 124 (6-[Prob.27])

Mean, 108 (5-[Ex.7])
Measurements, table of, 128–129
Median, 108 (5-[Ex.7])
Money, 132
Multiples, 3 (1-[Ex.2]), 87 (4-[Ex.11]),
  88 (4-[Ex.12]), 90 (4-[Ex.13]), 91
  (4-[Prob.3]), 94 (4-[Prob.3]), 112
  (5-[Prob.12]), 117 (5-[Prob.12])

Odd-even, 85 (4-[Ex.8]), 86 (4-[Ex.9]),
  86 (4-[Ex.10]), 92 (4-[Prob.8]), 100
  (4-[Prob.8]), 105 (5-[Ex.1]), 125
  (6-[Prob.31])
  properties, 85, 87
Opposites, 113 (5-[Prob.2])

Parallel lines, 36 (2-[Prob.10]), 45
  (2-[Prob.10]), 134
  definition, 133
Percentages, 7 (1-[Ex.8]), 23 (2-[Ex.3]),
  26 (2-[Ex.6]), 35 (2-[Prob.3]), 39
  (2-[Prob.3]), 58 (3-[Ex.8]), 60
  (3-[Ex.9]), 62 (3-[Prob.6]), 67
  (3-[Prob.6]), 105 (5-[Ex.3]), 111

168